Make: Raspberry Pi and AVR Projects

Cefn Hoile, Clare Bowman,
Sjoerd Dirk Meijer,
Brian Corteil, Lauren Orsini

Make: Raspberry Pi and AVR Projects

by Cefn Hoile, Clare Bowman, Sjoerd Dirk Meijer, Brian Corteil, and Lauren Orsini

Printed in the United States of America.

Published by Maker Media, Inc., 1005 Gravenstein Highway North, Sebastopol, CA 95472.

Maker Media books may be purchased for educational, business, or sales promotional use. Online editions are also available for most titles (*http://safaribooksonline.com*). For more information, contact our corporate/institutional sales department: 800-998-9938 or *corporate@oreilly.com*.

Editor: Patrick Di Justo
Production Editor: Melanie Yarbrough
Copyeditor: Charles Roumeliotis
Proofreader: Kim Cofer
Indexer: Angela Howard
Interior Designer: David Futato
Cover Designer: Brian Jepson
Illustrators: Cefn Hoile, Clare Bowman, Sjoerd Dirk Meijer, Brian Corteil, and Rebecca Demarest
Cover Photographer: Sjoerd Dirk Meijer

November 2014: First Edition

Revision History for the First Edition

2014-11-04: First Release

See *http://oreilly.com/catalog/errata.csp?isbn=9781457186240* for release details.

978-1-457-18624-0

[LSI]

Contents

Preface

All of the projects in this book use the Raspberry Pi. Read along to discover how to receive Twitter alerts through a lamp, collect temperature data from a fish tank, control a game with Raspberries (the actual fruit!), or even learn how to make your own wearable one man band.

Book-wide bill of materials

Table P-1. *Book-wide bill of materials*

Item	Item
Raspberry Pi (Models B or B+)	Pi Cobbler
2x ADXL 345 triple axis accelerometer breakout board (labeled GY-291)	2x 100 ohm resistor
2x piezoelectric transducer	7x 170 point mini breadboards
2x Bluetooth UART transceiver (board name JY-MCU, running HC-06 firmware)	2x 1M ohm resistor
2x USB to UART serial programmer (CP2102)	3x 40-pin male header strip
2x ATmega328P-PU (flashed with Arduino Uno bootloader)	2x 5mm red ultrabright LED

Item	Item
6x 10k ohm resistors	4x 22pF ceramic capacitor
4x 100nF ceramic capacitor	Solid core wire
2x 4-way female ribbon cables	2x LiPo battery and charger or 6x AAA battery pack and batteries
Jumper wires male-male	Jumper wires male-female
DS18B20 digital temperature sensor	Push-to-make button/switch
3mm acrylic base plate	2x 2.5mm hex bolts, length 12mm
3mm acrylic top plate	2x 2.5mm nuts
4x PCB hex spacers 25mm	16x 3mm hex bolts, length 6mm
4x PCB hex spacers 10mm	4x sticky rubber feet
2x PCB round spacers 6mm	Double-sided sticky foam tape/pads
Wireless USB adapter	4.7k (1/4 Watt) resistor
Darlington array ULN2803A	Spotlight Kit LED RGB matrix
1x 400-point miniature breadboard	2mm heat shrink tube
Sparkfun Breadboard Power Supply 5V/3.3V part number PRT-00114	12V DC 2A power supply to 9V DC 2A power supply
1x 4-pole screw connector block for stripboard, w/ 0.1" pitch hole spacing	Wire 22 AWG
Stripboard prototyping PCB, w/ 0.1" pitch hole spacing	2.1 jack socket (breadboard compatible)
8GB SD card or 8GB SD card with NOOBS preloaded	2x button
1x ATmega328P-PU (DIP28) or ATmega8 (DIP28)	1x 10 uF capacitor
3x 100 nF/0.1 uF capacitor	2x 22 pF capacitor
16x 22 MΩ or 20 MΩ resistor (1/4W)	2x 10 kΩ resistor (1/4W)
1x 2.2 kΩ resistor (1/4W)	2x 68 Ω resistor (1/4W)
5x 330 Ω resistor (1/4W)	5x LED (5mm) (4 different colors)

Item	Item
1x 1N4148 diode	2x 3.6V Zener diode (max. 0.5W!)
1x 16 MHz crystal	

Table P-2. *Optional materials*

2x Copper stripboard 93 x 54mm	2x 28 pin DIP IC Socket Adapter
2x 40 pin male right angle header strip	ESD Anti Static Wrist Strap
1x stripboard (94 x 55 mm)	1x 28 pin DIP IC socket
Alligator clips	Small breadboard
1x 150 ohm resistor 1/4 watt through the hole	3x super bright green LEDs 20mA
1 x 315 ohm resistor 1/4 watt through the hole	3x super bright red LEDs 20mA
1 x 105 ohm resistor 1/4 watt through the hole	3x super bright blue LEDs 20mA

Table P-3. *Tools list*

Long-nosed pliers	Wire cutter
Wire stripper	Hot glue gun and glue sticks
Solder	Dedicated spot cutter or 4mm drill bit
Stripboard	Junior hacksaw
Soldering iron and holder	Permanent pen
Blue-tack (poster putty)	Ruler
G-clamp	Multimeter
Small vise	Saw
Hand drill	Hammer
Allen key to suit bolts	Stripboard cutter
Desoldering braid	Safety glasses
Solder sucker	Screwdrivers to suit
Craft knife	Masking tape

Conventions Used in This Book

The following typographical conventions are used in this book:

Italic
: Indicates new terms, URLs, email addresses, filenames, and file extensions.

`Constant width`
: Used for program listings, as well as within paragraphs to refer to program elements such as variable or function names, databases, data types, environment variables, statements, and keywords.

`Constant width bold`
: Shows commands or other text that should be typed literally by the user.

`Constant width italic`
: Shows text that should be replaced with user-supplied values or by values determined by context.

This element signifies a tip, suggestion, or general note.

This element indicates a warning or caution.

Using Code Examples

This book is here to help you get your job done. In general, you may use the code in this book in your programs and documentation. You do not need to contact us for permission unless you're reproducing a significant portion of the code. For example, writing a program that uses several chunks of code from this book does not require permission. Selling or distributing a CD-ROM of examples from Make: books does require permission. Answering a question by citing this book and quoting example code does not require permission. Incorporating a significant amount of example code from this book into your product's documentation does require permission.

We appreciate, but do not require, attribution. An attribution usually includes the title, author, publisher, and ISBN. For example: "*Make: Raspberry Pi and AVR Projects* by Clare Bowman, Brian Corteil, Cefn Hoile, Sjoerd Dirk Meijer, and Lauren Orsini (Maker Media). Copyright 2014, 978-1-4571-8624-0."

If you feel your use of code examples falls outside fair use or the permission given here, feel free to contact us at *bookpermissions@makermedia.com*.

Safari® Books Online

Safari Books Online is an on-demand digital library that delivers expert content in both book and video form from the world's leading authors in technology and business.

Technology professionals, software developers, web designers, and business and creative professionals use Safari Books Online as their primary resource for research, problem solving, learning, and certification training.

Safari Books Online offers a range of product mixes and pricing programs for organizations, government agencies, and individuals. Subscribers have access to thousands of books, training videos, and prepublication manuscripts in one fully searchable database from publishers like Maker Media, O'Reilly Media, Prentice Hall Professional, Addison-Wesley Professional, Microsoft Press, Sams, Que, Peachpit Press, Focal Press, Cisco Press, John Wiley & Sons, Syngress, Morgan Kaufmann, IBM Redbooks, Packt, Adobe Press, FT Press, Apress, Manning, New Riders, McGraw-Hill, Jones & Bartlett, Course Technology, and dozens more. For more information about Safari Books Online, please visit us online.

How to Contact Us

Please address comments and questions concerning this book to the publisher:

Make:
1005 Gravenstein Highway North
Sebastopol, CA 95472
800-998-9938 (in the United States and Canada)
707-829-0515 (international or local)
707-829-0104 (fax)

Make: unites, inspires, informs, and entertains a growing community of resourceful people who undertake amazing projects in their backyards, basements, and garages. Make: celebrates your right to tweak, hack, and bend any technology to your will. The Make: audience continues to be a growing culture and community that believes in bettering ourselves, our environment, our educational system—our entire world. This is much more than an audience, it's a worldwide movement that Make: is leading—we call it the Maker Movement.

For more information about Make:, visit us online:

Make: magazine: *http://makezine.com/magazine/*
Maker Faire: *http://makerfaire.com*
Makezine.com: *http://makezine.com*
Maker Shed: *http://makershed.com*

We have a web page for this book, where we list errata, examples, and any additional information. You can access this page at: *http://bit.ly/make-raspberry-pi-and-avr-projects*.

To comment or ask technical questions about this book, send email to: *bookquestions@oreilly.com*.

About the Authors

Clare Bowman enjoys hacking playful interactive installations and co-designing digitally fabricated consumer products. She has exhibited projects at Maker Faire UK, Victoria and Albert Museum, FutureEverything, and Curiosity Collective gallery shows. Some recent work includes "Sands Everything," an interactive hourglass installation interpreting Shakespeare's Seven Ages of Man soliloquy through gravity-controlled animated grains, and more.

Brian Corteil has never grown up, and still loves playing with computers, micro electronics, Lego, and video games. His first computers were a ZX80 then a TI-99, and finally an Acorn Electron. He is one of the founding members of Makespace, the place to make, fix, break stuff, and meet great people in Cambridge.

Cefn Hoile sculpts open source hardware and software, and helps others do the same. After 10 years of industrial R&D, Cefn founded *www.shrimping.it* to help schools and hobbyists source and adopt electronics prototyping materials in the classroom. One of the project's freely licensed builds, "The Shrimp," is an Arduino-compatible breadboard layout on which the ShrimpKey and Picussion projects in the book are based. Cefn is currently completing a PhD in Digital Innovation at Highwire, University of Lancaster, UK.

Sjoerd Dirk Meijer is the maker of ShrimpKey (DIY MakeyMakey) and a Scratch programming educator. He is also interested in (primary) education, giftedness, and making/Maker Ed. He can be found on Twitter @fromScratchEd.

Lauren Orsini is a technology journalist in Washington, DC. She writes about developer issues, tech education, and DIY hardware hacking for ReadWrite. Her new book, *Otaku Journalism: A Guide to Geek Reporting in the Digital Age*, is a new media journalism handbook to navigating Internet-age reporting.

Acknowledgments

Clare Bowman and Cefn Hoile

At OSHCamp (Open Source Hardware Camp) in Hebden Bridge, UK, we got to chatting with a man whose daughter had hyposensitive autism, a sensory problem where she is under-sensitive to stimuli. She found it difficult making friends as she did not know how to judge the personal space that others expected, and wanted to understand when she was getting too close.

We discussed a wearable device that would help her identify this personal space, loosely involving an ultrasonic module and vibration motor, housed in a necklace or belt that his daughter could design with a jewelry maker, so it was aesthetically pleasing and personal to her. Sadly we lost touch, and never found out if he managed to get the device built (if you are reading this book, we'd love to hear from you).

Through this encounter we started thinking about wearables, and at a body percussion workshop soon afterward, we realized how wireless wearable technology could turn your body into a drum machine, and Picussion was born. The UK Raspberry Jam network introduced us to the Pi, which is perfect for this emerging application, and the design began to take shape. Fast forward a few years, after a baby and a lot of experimentation, and we have included it in this book. Hope you enjoy building the project and your noisy journey with Picussion!

We want to thank the following people who have been immensely supportive while working on this project.

Firstly, Liz Edwards, a PhD student at Highwire, Lancaster University, who has been a keen contributor to Chapter 2, and her design skills have been invaluable. Thank you so much for your support and friendship. Paul Bowman, an inventor, cracking photographer, and wonderfully talented man in many other ways. Rachel Diss, an inspirational and supportive occupational therapist. The fabulous One Sam Band who performs around the North West of England. Our daughter Sky has put up with both parents hunched over laptops and sprawling circuitry while writing this chapter, so thank you Sky for being amazing, being our human guinea pig, and making us laugh when we were feeling tired and squinty-eyed. @ShrimpingIt thanks Fritzing.org for the use of vector graphics elements in the Picussion and ShrimpKey build diagrams. Thanks to Manu Brueggemann for camerawork and hand jiving. Also thanks to all our friends who have helped look after Sky and fed us cake while writing this chapter.

Brian Corteil

I first heard about the Raspberry Pi about three years ago, when a video about the prototype went viral. Back then I saw a small PC that would be cheap enough to leave in my projects, allowing them to connect to the Internet at below the cost of an Arduino Uno and Ethernet shield. I followed the project, waiting for the day that it would be released. When the day came, the release of the Raspberry Pi took down two of the UK's major electronic component suppliers' websites, but I was one of the lucky few who managed to order one. As of the writing of this book, the Raspberry Pi has sold over three million units!

I have a thing about pretty lights; RGB LEDs are candy to me. I love the way light can shape and bring out the detail in objects. So when I went to my local B&Q store, a child's night light caught my eye. The night light was in the shape of a giraffe, and my wife Sarah loves giraffes. She has a collection of at least 200, so as I turned the lamp over, I started formulating a project. I wanted a lamp that would create a "bat signal" for my wife to contact me when it flashed, since she has a habit of leaving her mobile phone in her purse with her coat thrown on top.

There are a number of people I would like to thank for their help and support.

First to the most important people in my life, Sarah, Billy, and Tom. Sarah, thank you for being the inspiration for this project and putting up with me as I have been writing this book when I should be doing stuff around the house. Billy and Tom, thank you both for letting me work on this project and your understanding, when I should have been playing with you both. It is my greatest joy watching you both grow up, I know you're going to be fine men that I will be proud to say are my sons.

I would like to thank the past and present members of Makespace for their support, help, and encouragement in the building and coding of this project. I wouldn't have been able to finish this project without them.

Sjoerd Dirk Meijer

Computers aren't just tools that support our daily lives, they're 21st century tinkering materials; you can make them do what you want (and not the other way around). But for that we have to learn the possibilities of computers and computer programming.

That's why I wanted to teach my daughter and son programming and, after searching the Internet, I found the graphical programming environment Scratch (*http://fse.link/r04*). It was exactly what I was looking for.

Most programming languages are English-based; so children in non-English countries have a disadvantage. They have to learn to read and translate the programming language, learn the syntax of the programming language, and have to learn to think as a programmer.

With Scratch they only have to learn to think as as programmer, because the code is on blocks and the code is translated into many languages.

Making computer programs is fun. But it's even more fun if you can control your program with a homemade controller. After viewing Jay Silver's TED talk "Hack a banana, make a keyboard" (*http://fse.link/r05*) I knew this was the "tool" I was looking for. After playing with it, I decided to use the MaKey MaKey in the programming workshops I host. But I needed more than one. Unfortunately, the MaKey MaKey isn't cheap and buying 10 of them would be too expensive for me. In the same period I found the Shrimp. After reading that the MaKey MaKey is Arduino-based, I was certain that it would be possible to make my own (cheaper) "MaKey MaKey" with a Shrimp. And it was!

I already owned a Raspberry Pi. And even though I love this credit card–sized computer, I couldn't find a functional way for me to use it. Until I found Simon Walter's ScratchGPIO (*http://fse.link/r06*). It's a special version of Scratch on the Raspberry Pi that can control the GPIO pins. With ScratchGPIO the Raspberry Pi can, like the Arduino, be used to control LEDs, motors, etc. Now children (and grown-ups) can program their own physical projects, without using "difficult" programming languages, like Processing, Python, and C.

Special thanks go to:

Jay Silver and Eric Rosenbaum for making the MaKey MaKey and licensing it as open source and open hardware. Without them I wouldn't have made the ShrimpKey.

Cefn Hoile and Stephan Baerwolf for their great help on the ShrimpKey source code.

And, most importantly, a big thank you to my wife and kids. They had to cope with a husband/father who used all of his spare time to write and therefore couldn't give them the attention they deserve. You three are amazing!

Lauren Orsini

I am not a programmer, or at least I wasn't until the Raspberry Pi came along. I wouldn't dare do anything that'd risk breaking my pricey Mac-

Book. I always wanted to learn how to work with hardware, I just never had the right tools.

Enter the Raspberry Pi, which took away all of my self-imposed barriers and excuses. It was too inexpensive not to try out. And once I started messing with it, I couldn't stop. The best part, of course, is the enormous community of amateurs and experts alike, all chatting on the Web about their projects, concerns, and solutions for Raspberry Pi development.

By contributing here, I am glad to give back to the community that made me love Raspberry Pi. I hope you pick up this book and—whether you are a beginner or seasoned coder, an engineer, developer, or a writer like me—catch the Raspberry Pi bug.

Special thanks go to:

My husband, John, was a huge part of this project. While I am an amateur, John has some background in object-oriented programming and helped me to understand and customize the syntax behind the commands this project requires. And while I definitely got frustrated at times, John never gave up.

My friend, Patches, a professional programmer who was invaluable to me during my first attempt ever at tackling MySQL. Patches also turned me on to D3JS so I wouldn't have to try to build a database-reading graph from scratch.

My mom, who dedicated her first published book to me right after I was born. Sorry it took 27 years for me to return the favor, Mom!

And of course, my betta fish. I originally designed this project for Fintan, an older blue-and-orange fish, and it was originally named after him. Fintan didn't live to see this section published, but you can see photos of him peppered throughout it. He had some big shoes to fill so I brought home Levi, a juvenile betta with dark coloring, to personally test the final draft of this project.

And special thanks goes to Nelson Neves for his review of this book.

1 Introduction

Eben Upton invented the Raspberry Pi. As a teacher, he noticed that engineering students were getting worse, not better, at basic computer science concepts. He realized that students were not tinkering with the guts of their computers the way he had with his childhood Amiga. His solution was to prototype a $35 hackable computer, which we now know as the Raspberry Pi. With something so cheap and simple to program, engineering students could more easily study computing concepts.

Millions of people outside of his classroom scrambled to get their hands on a Pi. What he had designed for his students was now adopted by makers, tinkerers, and creators all over the world.

The Power of the Pi

The Raspberry Pi is a small computer. It has Ethernet networking, USB ports for peripherals, HDMI and composite video outputs, and a headphone jack for audio. It also has a small number of General Purpose Input/Output (GPIO) pins that can be controlled with software to send and receive binary information to control electronic circuits and microcontrollers.

The Pi boots from an SD card that contains its operating system and programs. The "New Out Of Box System" (NOOBS) preconfigured SD card gives new users a simple user interface for installing various operating systems, including the Raspbian Linux distribution, recommended by the Raspberry Pi foundation.

At the time of this writing, the Raspberry Pi Foundation released a new Pi model called the B+. This model offers more GPIO pins, additional USB ports, better audio, and lower power consumption. All four projects that are detailed in this book work with the B+.

Raspbian

Raspbian Linux makes the Raspberry Pi's desktop visible using an HDMI or RCA monitor, though most people prefer HDMI for a pixel-perfect display. While the Pi itself is cheap, the cost of peripherals like a monitor, keyboard, mouse, and power supply will cost many times more than the device itself.

One way to dodge such extra purchases is to network your Pi. All you need are a standard USB power cable and an Ethernet cable. Plug the USB into a power source, and plug the Ethernet into a router (or a laptop/desktop computer that is configured for network sharing), and you're good to go. Once the Pi has booted, you can see its desktop using a remote access program like VNC. With this type of setup, you can use your Pi without a keyboard, mouse, or monitor—"headless," as it is known in the Raspberry Pi community. However, because it sends the desktop as a series of images over VNC, it is not appropriate for high-frame-rate applications such as video playback.

Who Is This Book For?

This is not a beginner's book, but effort has been made to walk even novice readers through the projects. An electronics and soldering background would be nice to have prior to starting these projects, but is not necessary.

Projects in This Book

Five makers from all over the world have come together to share a few of their projects and inspire readers. This book consists of four main projects: Picussion, Raspberries from Scratch, Internet of Fish, and a Giraffe Mood Lamp.

Picussion, in Chapter 2, is a wireless, wearable device, made by Clare Bowman and Cefn Hoile, that can be configured by a Raspberry Pi to trigger sound samples by knocking/hitting it or orienting it in space.

The **Raspberries from Scratch** in Chapter 3, made by Sjoerd Dirk Meijer, turns everyday, conductive materials into a homemade conductive keyboard. This project walks you through building the conductive keyboard, and then walks you through making two computer programs with the Scratch language.

Chapter 4's **Internet of Fish**, by Lauren Orsini, is a smart, communicative thermometer. With this project, your Pi texts your cell phone with any overheating or cooling emergencies, and it also feeds a local web server that reads and visualizes the thermometer's data. This project could be used in other ways, such as monitoring a swimming pool, or maintaining a perfect temperature for an at-home brewery.

The **Giraffe Mood Lamp** in Chapter 5 has the brains of a Raspberry Pi and the exterior of a lamp. This project, created by Brian Corteil, is able to receive Twitter messages by harnessing the Twitter API, and flashes when it finds a new message.

2 Picussion

By Clare Bowman and Cefn Hoile

Have you ever dreamed of playing in a one-man band? Follow along and construct cheap, wearable devices, using them to trigger interactive audio running on a Raspberry Pi. We call this project (Pi)cussion (Figure 2-1).

Figure 2-1. *Picussion*

Although the wearables themselves are based on an ATmega chip, the Raspberry Pi is integral to the Picussion project. The Pi runs a full desktop Linux OS based on Debian with easy-to-install software packages. It is simple to configure the hardware, tools, and software libraries needed to create Picussion. With a few lines of Python code, you can use the Pi to trigger communication using a Bluetooth USB adapter, and to play audio samples to headphones or speakers.

During this chapter you will:

- Use the Arduino IDE to upload Arduino code to the wearable.
- Execute Python, running on the Raspberry Pi to:
  - Connect to the wearables
  - Process the sensor data
  - Control remote behavior and local multimedia
- Walk through five exercises to build the final one-man band Picussion.

Exercise 1: Blink
: Build a DIY microcontroller circuit and test it using the Raspberry Pi, proving that your circuit is alive and can be programmed.

Exercise 2: Knock
: Establish that Python can be used to read tap sensor information from the microcontroller over wires, using a serial protocol.

Exercise 3: Knock Wireless
: Replicate the Knock behavior over a Bluetooth wireless serial link.

Exercise 4: Talking Wearables
: Show how multiple sensors can each trigger their own audio sample.

Exercise 5: One Man Band
: Go beyond hitting the sensors, to detecting complex gestures that trigger different instruments.

Let's get building!

Materials and Tools

Before you begin, you will need a Raspberry Pi B (or a B+), a keyboard, mouse, monitor, and an SD card. In addition, you need a USB 2.0 4-way hub (unless you are using the B+ model), and a Bluetooth dongle, since the Pi doesn't have built-in Bluetooth.

There is a community resource listing Bluetooth dongles that have been tested with Raspbian at eLinux (*http://bit.ly/1xTiTo6*).

Solderless Breadboarding Materials

The materials you'll need to build a Picussion wearable module on a solderless breadboard are listed in Table 2-1.

Table 2-1. *Bill of materials*

Item	Item
ADXL 345 triple axis accelerometer breakout board (labeled GY-291)	100 ohm resistor
Piezoelectric transducer	2x 170 point mini breadboards
Bluetooth UART transceiver (board name JY-MCU, running HC-06 firmware)	1x 1M ohm resistor
USB to UART serial programmer (CP2102)	40-pin male header strip
ATmega328P-PU (flashed with Arduino Uno bootloader)	5mm red ultrabright LED
3x 10k ohm resistors	2x 22pF ceramic capacitor
2x 100nF ceramic capacitor	Solid core wire
4-way female ribbon cables	LiPo battery and charger or 3 x AAA battery pack and batteries

Solderless Breadboard Tools

These tools are helpful to make a breadboard build more pleasant, although you can probably improvise with other tools if used carefully (see Figure 2-2 and Table 2-2).

Figure 2-2. *Solderless breadboard tools*

Table 2-2. *Solderless breadboard tools*

Long-nosed pliers
Wire cutter
Wire stripper

Double Components

This bill of materials lists the required components to build one wearable, but if you are interested in creating the full One Man Band project you will need to double this list so that you can build two wearables.

Soldering Materials

Soldering your wearables—which is optional—will make them more robust, and free up breadboards for your next project; prototyping on breadboard is adequate for many applications. To solder your materials together, use the materials in Table 2-3.

Table 2-3. *Soldering materials*

Soldering materials
Copper stripboard 93 x 54mm
28-pin DIP IC socket adapter
40-pin male right angle header strip

Soldering Tools

Make sure you have the materials shown in Figure 2-3 and listed in Table 2-4 in hand if you plan to solder components onto copper stripboard. A good soldering iron is especially important. A G-clamp can secure your stripboard to a stable surface when cutting the board with a hacksaw.

Table 2-4. *Soldering tools*

Soldering tools
Solder
Dedicated spot cutter or 4mm drill bit
Stripboard
Junior hacksaw
Soldering iron and holder
Permanent pen
Soldering stand (helping hand)
Ruler
G-clamp
Voltage meter
Toothbrush
Tea bags
Your favorite biscuits

Figure 2-3. *Soldering tools*

Configuring Your Raspberry Pi

In this section you will add additional software to the Raspberry Pi. If your Pi is networked and you have Raspbian running, double-click the LXTerminal (see Appendix A for a Raspbian installation) to bring up a console window. You will type commands into this window to build the projects in this chapter.

Type the following lines into the console, and press Enter after each line. The `sudo` command means that these commands are run as *super user*, giving the commands full privileges to change your Raspberry Pi system files and configuration.

```
sudo apt-get update
sudo apt-get install arduino python-serial python-numpy python-tk
sudo apt-get install python-bluez bluez blueman firmware-atheros
```

You should now see text scrolling down the screen. This text shows the process of downloading information from the online software repositories provided as part of Raspbian. You may be asked to confirm certain changes.

A USB-to-UART serial adapter works with the Arduino IDE in Raspbian to send programs over to the tiny computer running inside the wearables that you will soon be building. To give the Pi user account privileges, they must be added to the dialout group. Type the following into the LXTerminal console window:

```
sudo usermod -a -G dialout pi
```

Once this is completed, save anything else you're working on. Click the Start menu, and choose Logout. Then log back in again. The default username and password are pi and raspberry, respectively. New privileges are loaded into your user account at login, so they should then be available when you run the Arduino software.

Restarting the Pi is an alternative, even though it actually takes longer. To do this, you can run the following command to instantly restart the Pi (close any important programs before running this, as they will be force-closed):

```
sudo shutdown -r now
```

Picussion Source Code

Before starting the first Exercise, be sure and download the source code for this project. The source code is on GitHub (*https://github.com/backstopmedia/maker/*), but see Appendix A for instructions on downloading the Picussion scripts to the Rasberry Pi desktop.

Exercise 1: Blink

In this section, we show you how to build a DIY microcontroller circuit from scratch (see Figures 2-4 and 2-5), and upload your first code to control it.

Figure 2-4. *Completed Picussion wearable module*

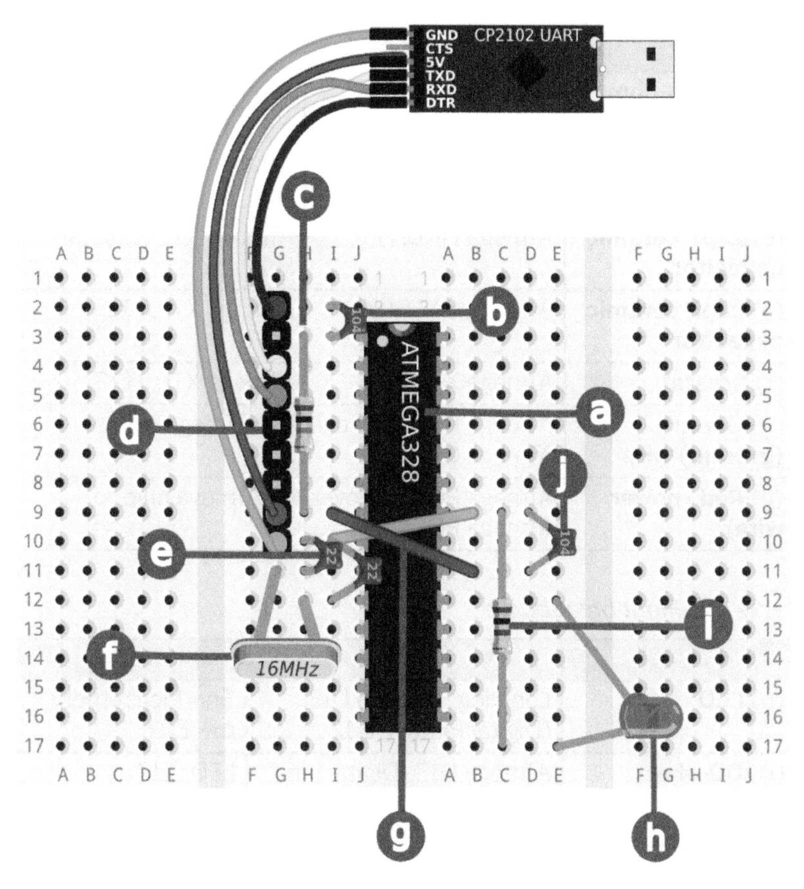

Figure 2-5. *The complete DIY microcontroller circuit*

Use Tables 2-5 and 2-6, along with the detailed list of components with details and tips to attach all of your components into the proper holes of your breadboard.

Table 2-5. *Left breadboard*

Component	From	To
(a) **ATmega328**	Place the chip straddling the two breadboards with the half-moon shape at the top, and two rows left empty above	
(b) **104pF ceramic capacitor**	ATmega RST (Reset), I:3	One hole above ATmega RST, I:2

Component	From	To
(c) **10 kilo-ohm resistor**	ATmega RST (Reset), H:3	ATmega POW (Power), H:9
(d) **9-pin male header**	G:2	ATmega GND (Ground), G:10
(e) **22pF ceramic capacitor**	ATmega GND, H:10	ATmega XTL (Crystal), H:11
(e) **22pF ceramic capacitor**	ATmega GND, I:10	ATmega XTL, I:12
(f) **Crystal**	ATmega XTL, G:11	ATmega XTL, H:12
(g) **Green (ground) wire**	ATmega GND, I:10 over the ATmega chip to ATmega GND, B:9	
(g) **Red (power wire)**	ATmega POW, I:9 over the ATmega chip to ATmega POW, B:11	

Table 2-6. *Right breadboard*

Component	From	To
(h) **LED**	Long leg, ATmega D13 (Digital Pin 13), E:12	Short leg, empty row E:17
(i) **100-ohm resistor**	ATmega D13 (Digital Pin 13), C:9	LED C:17
(j) **104pF ceramic capacitor**	ATmega GND, D:9	ATmega POW, D:11

(a) ATmega microcontroller

The ATmega microcontroller is a black oblong integrated circuit with numbers printed on it, with 28 silver legs, or pins. It is the computer at the heart of your circuit, and has inputs and outputs, which can sense or trigger things in the outside world.

Solderless Breadboarding Tips

Circuits look neater and are less likely to have wires pull out or touch each other when components are trimmed to length and appear flush with the board. Before cutting the wires:

- Add a bend to a connection before insertion (looks like a "C" from the side)
- Allow extra length for the wire to insert into the breadboard
- Think twice, since you may not be able to reuse this component in a different layout in the future

Placing the ATmega

You can break the legs if you force the ATmega into your solderless breadboard. If the legs are not at right angles to the ATmega, ease them into position by gently pressing one side of the ATmega against the tabletop (14 pins at a time). Carefully align the ATmega, checking that the half-moon shape is at the top, with two empty breadboard rows above the chip. Once the legs are lined up well on the correct holes, press down softly to slide them in. Then, push down hard to ensure a good connection. When the ATmega slides fully into the board, there is about 3mm of movement.

(b) Ceramic capacitor

Look for the 100 nF ceramic capacitor. This is a small disc with two thin wires coming out of it. This decoupling capacitor smooths electrical spikes, so that the incredibly brief "reboot" signal sent through to Pin 1 is stable and reliably detected.

(c) 10 kilo-ohm pull-up resistor

A positive voltage, usually 5V, "pulls up" the reboot pin (Pin 1) through this resistor. The ATmega will keep running as long as it gets a high voltage on this pin. When reprogramming, the brown wire is briefly connected directly to 0V (a stronger signal than the flow from 5V limited by the resistor). Pin 1 therefore gets pulled to 0V, which causes the chip to reboot. After rebooting, the ATmega listens for a new program sent over the orange wire. If nothing is sent, it runs the last program.

(d) 9-pin power and programming header

A series of copper pins will be used to program and provide power to the ATmega microcontroller. Keep the strip intact as you slide the nine pins in. Not all of the nine pins will be used in this circuit, but having them all in the right place helps you position everything else.

(e) 22 nanofarad ceramic capacitors

These help to ensure the crystal circuit resonates in a reliable way.

(f) 16 MHz crystal

A computer is a bit like clockwork. The first ever digital computer was built using clock-making techniques, with cogs representing numbers! This quartz crystal acts like the clock's pendulum, causing the mechanism to tick along. The 16.000 indicates the number of back-and-forth movements this crystal generates per second, in megahertz. One hertz means once per second, and one megahertz means one million times per second.

(g) Power and ground wires

The ATmega chip is broken up internally into separate parts, each of which needs a stable power supply. Two wires are needed to connect power and ground across to the correct legs on the righthand half of the ATmega.

(h) Light emitting diode (LED)

A diode allows electrical current to easily flow in only one direction. The LED has one positive, and one negative. The longer lead is positive and the shorter lead is negative.

(i) 100 ohm current-limiting series resistor

The circuit will be running at between 3.7 and 5 volts, and LEDs are rated around 1 to 2 volts. This resistor is connected in series with the LED, limiting the amount of electrical current flowing to prevent the LED from overheating and being destroyed. Some of the voltage will be dropped by the resistor, and only a suitable share of it is applied across the LED.

(j) 100 nanofarad ceramic decoupling capacitor

This capacitor smooths the power supply voltage available to the ATmega when sudden drops in voltage are caused by demands from other components in a circuit, such as motors or LEDs.

Attaching the UART

Now you need to attach the serial UART (CP2102) to the breadboard. Push the colored wires onto the pin headers as shown in Figure 2-6.

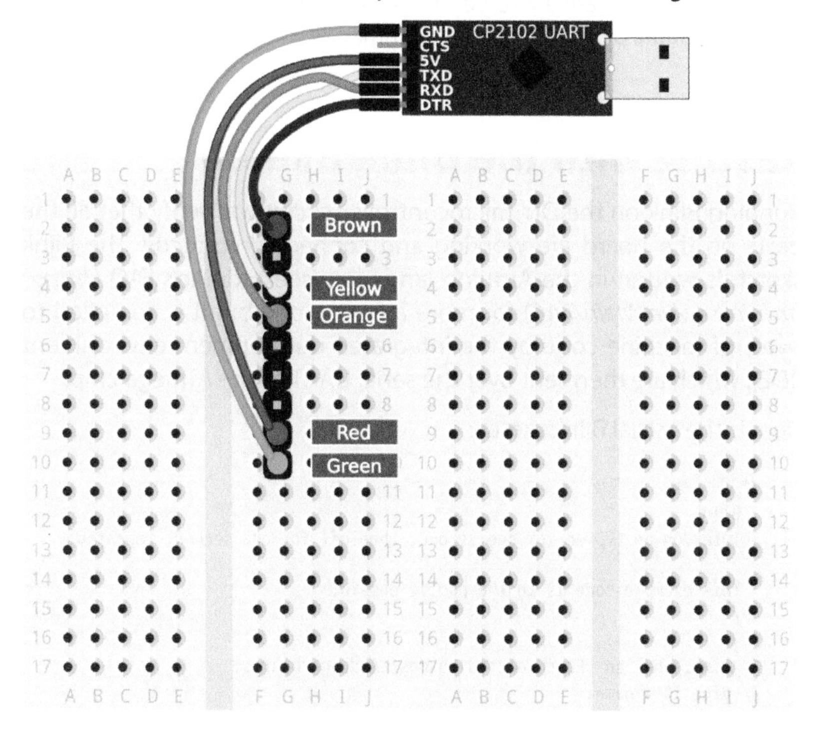

Figure 2-6. *Plugging the UART into the board*

Now that you have built your Arduino-compatible DIY microcontroller breadboard, it's time to test your hardware and make sure it's working properly. If you've configured everything correctly, you are setting the stage to light up the LED.

Serial UART

A serial UART connection provides a standardized way for two computers to talk to each other using just two wires. Information is sent over the wires as a series of bits. A bit is a single value that can have have two states (one or zero/on or off). These values are sent simply as high and low voltages with an agreed timing gap between each value. The bits are grouped into bytes (chunks of 8 bits) that can be used to represent a series of 8-bit numbers (from 0 to 255). These bytes can alternatively represent written alphanumeric characters, punctuation, or simple printing commands.

Testing with the Blink Sketch

Running Blink on the DIY microcontroller circuit will verify that all the parts on the board are working and connected correctly. The Blink sketch is written in the Arduino language (see Arduino's FAQ (*http://arduino.cc/en/Main/FAQ*) for more information), but it is compiled to bytes of machine code by the integrated development environment (IDE), which are then sent over the serial UART to the ATmega chip.

Here is the actual Blink sketch:

```
/*
  Blink
  Turns on an LED on for one second, then off for one second, repeatedly.

  This example code is in the public domain.
 */

// Pin 13 has an LED connected on most Arduino boards.
// give it a name:
int led = 13;

// the setup routine runs once when you press reset:
void setup() {
  // initialize the digital pin as an output.
  pinMode(led, OUTPUT);
}

// the loop routine runs over and over again forever:
void loop() {
  digitalWrite(led, HIGH);   // turn the LED on (HIGH is the voltage level)
  delay(1000);               // wait for a second
  digitalWrite(led, LOW);    // turn the LED off by making the voltage LOW
```

```
  delay(1000);                // wait for a second
}
```

You may see the LED flash once or twice as the circuit is powered up, and each time the serial connection is reset by the IDE when reprogramming your wearable. This is normal behavior, and indicates that the bootloader on the ATmega328 is starting up correctly.

Perform the following steps:

1. Plug the serial UART into the Raspberry Pi USB port.
2. Launch the Arduino IDE.
3. Open Blink from File → Examples → 01.Basics → Blink (see Figure 2-7).
4. Select Tools → Board, then select the Arduino Uno entry so that the IDE knows how to compile programs for it properly.
5. Select Tools → Serial Port, and click /dev/ttyUSB0.
6. Now you can upload the Blink sketch by choosing File → Upload.
7. The LED should flash briefly, and then the IDE should report the program was uploaded successfully in the message pane at the bottom of the window. The LED should start to flash, staying on for one second, then off for one second, until the power is cut.

Take a look at the instructions within the Blink sketch and see if you can make sense of how this behavior is implemented. Try changing the timing of the `delay()` function calls to modify the blinking regime of the circuit. After you make your changes, choose Save, then Upload in the IDE to send your first personalized program onto the wearable.

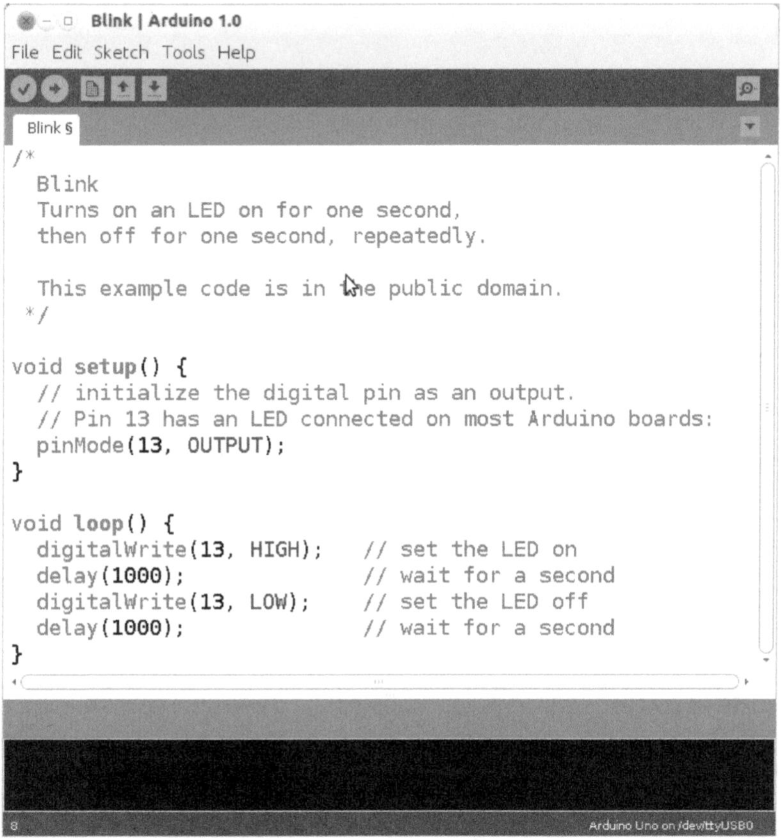

Blink | Arduino 1.0

File Edit Sketch Tools Help

Blink §

```
/*
  Blink
  Turns on an LED on for one second,
  then off for one second, repeatedly.

  This example code is in the public domain.
 */

void setup() {
  // initialize the digital pin as an output.
  // Pin 13 has an LED connected on most Arduino boards:
  pinMode(13, OUTPUT);
}

void loop() {
  digitalWrite(13, HIGH);   // set the LED on
  delay(1000);              // wait for a second
  digitalWrite(13, LOW);    // set the LED off
  delay(1000);              // wait for a second
}
```

8 Arduino Uno on /dev/ttyUSB0

Figure 2-7. *The Blink example, shown loaded in the Arduino IDE*

Troubleshooting

If you've successfully blinked your LED, congratulations. You can skip this section. If it isn't working, here are some things to consider:

- If you are seeing a *not in sync* error reported in the IDE, this means the IDE can't communicate with the ATmega328 bootloader to send a program. Assuming your chip has a bootloader on it, this means that there's an error in the wiring, such as the chip being inserted upside down, not being pushed down and seated properly, or some component or wire not attached in the right location.

- It's common to attach the wrong-colored wires to the wrong pins, even mistakenly connecting 5V to GND, directly short-circuiting the power supply. If the /dev/ttyUSB0 entry disappears from the list under Serial Ports, this is a bad sign. You've probably miswired the CP2102 USB UART module that is attached to the 9-pin male header. This can often prevent the UART from powering up properly. So double-check that the ribbon cable is attached properly to the DIY microcontroller board according to Table 2-7.

Table 2-7. *USB module wiring the ribbon cable to the Shrimp (note there are different conventions and the TX and RX may need to be swapped for this to work)*

Red	5V
Orange	RXD
Yellow	TXD
Green	GND
Brown	DTR

Robust Components

These components are robust. We have created circuits with extreme errors, and have not yet destroyed any of the core components of a DIY microcontroller board. While there's always a chance that the wrong wiring could fry a component, in our experience once the wiring error is fixed, we find the devices come back to life right away.

Exercise 2: Knock

The Blink sketch illustrated how the ATmega in our DIY microcontroller circuit can control outputs such as LEDs attached to its pins. It can also respond to inputs from attached sensors. In the simplest case, it can monitor the voltage on one of its pins and be programmed so that voltage changes trigger various behaviors.

Each small impact on your piezo (Figure 2-8) will get reported back to the Raspberry Pi, where richer behaviors can be triggered. When the board is powered up, it waits for a while to see if you're sending a new program over serial. If nothing is detected, it runs the last program you sent, which can use the serial link for real-time communication.

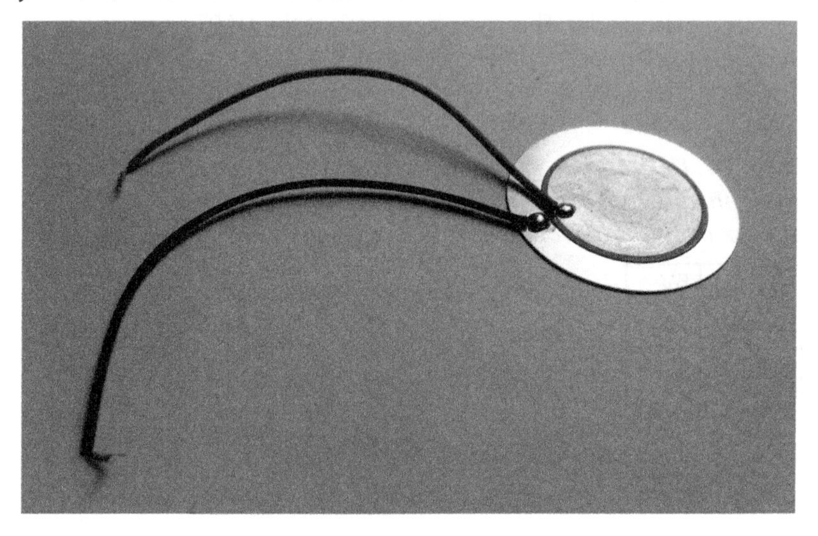

Figure 2-8. *A piezoelectric transducer element (unhoused)*

You now have a working DIY microcontroller circuit in front of you. We are using a 20mm low-profile piezo that doesn't have a plastic housing, since the exposed piezo provides us with increased sensitivity. Figure 2-9 shows how we add the piezo to the circuit and Table 2-8 outlines the components.

Piezos come presoldered, which is handy, but they can be delicate. Handle them gently and don't let any tiny children play with them, since they will only last seconds before falling apart.

Why Not Wear a Raspberry Pi?

You may be asking yourself, why create wearables? Why not just attach a Raspberry Pi to your body? This is certainly feasible, but for games with multiple distributed percussion sensors, it's overkill, adding complexity and reducing usability. For example, the AVR wearables boot and connect almost instantaneously, while a Pi can take much longer to launch a multimedia desktop. Pis need a relatively high-capacity stable power supply, where the AVR can run directly off cheap batteries for quite a long time. The cost of the AVR wearables is extremely low, compared to the equivalent deployed on a Pi (which would need something like an AVR for interfacing with the physical world anyway). This means you can afford to deploy a single wireless wearable for every limb!

Table 2-8. *Working on the right breadboard*

Component	From	To
(d) **Piezo element**	Ground, use black wire, I:10	Series resistor, use red wire, I:13
(a) **Blue wire**	ATmega AO1 (Analog Output 1), E:8	Piezo red wire, G:13
(b) **Green wire**	ATmega GND (Ground), E:9	Piezo black wire, G: 10
(c) **1 megaohm resistor**	Piezo black wire, H:10	Piezo red wire, H:13

(d) Piezo element

If the black and red wires of your piezo keep moving out of the breadboard, crimp a male pin to the wires for reliability. You will be attaching the piezo between Ground and Analog Pin 0 (zero), but routing via the blue wire (b) and green wire (c) so that you can easily place a large resistor (d) in parallel (see piezo details (*http://bit.ly/ZEf4at*) for more information).

(a) Blue wire

You are connecting this wire to analog input pin zero (0), a sensor pin that can detect a range of voltage levels between 0V and 5V (not just on or off like the digital input pins). Tapping harder or softer on the piezo will generate different voltage levels to the ATmega.

(b) Green wire

This gives us a connection back to ground, a reference voltage of 0V. The voltage on Analog Output 1 will return to this 0V reference after the voltage spike from tapping the piezo has dissipated.

(c) 1 megaohm resistor

A 1 megaohm resistor (a very large resistor) is connected alongside the piezo, allowing the small amount of energy generated by each tap to be discharged. This helps prevent overloading the ATmega input pin. It also means that any large spike of voltage from tapping the piezo doesn't linger for too long, so the circuit is ready to detect another tap soon after. It doesn't matter which direction the resistor is facing.

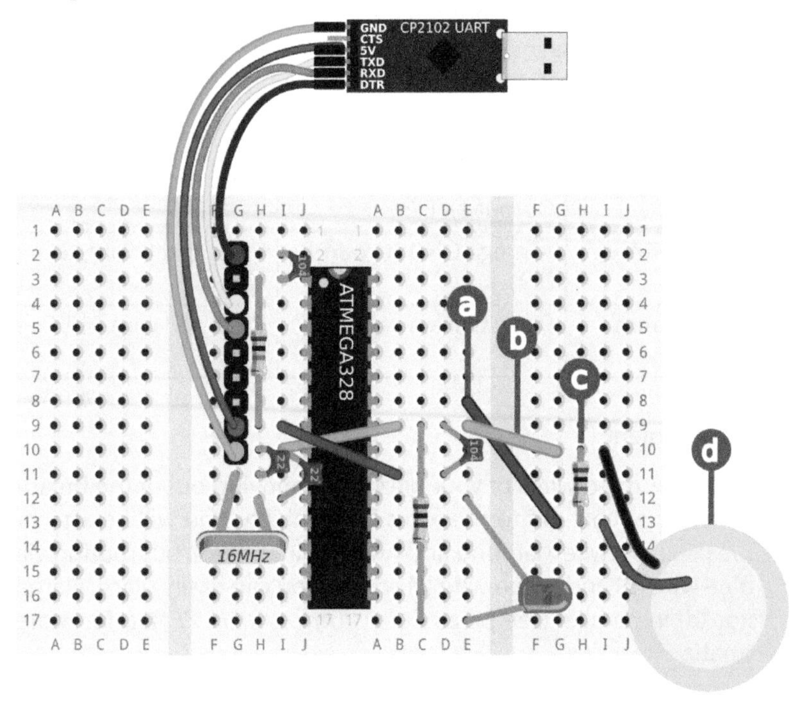

Figure 2-9. *Adding the piezo to the DIY microcontroller circuit*

Loading the Knock Sketch

Let's load the Knock sketch onto our wearable (DIY microcontroller circuit and piezo components):

1. Plug the UART from the DIY microcontroller circuit directly into your Raspberry Pi with the Raspbian GUI running. Be careful that your piezo doesn't come unplugged.
2. Launch the Arduino IDE.
3. Open the Knock sketch from File → Examples → 06.Sensors.
4. As with the earlier Blink build, check that the correct entries are selected under Tools → Serial and Tools → Board.
5. Click File → Upload, and check that the message pane shows a successful upload.

Here is the Knock sketch code:

```
const int ledPin = 13;       // led connected to digital pin 13
const int knockSensor = A0; // the piezo is connected to analog pin 0
const int threshold = 100;  // threshold value to decide when the
                            // detected sound is a knock or not


// these variables will change:
int sensorReading = 0;      // variable to store the value read
                            // from the sensor pin
int ledState = LOW;         // variable used to store the last LED
                            // status, to toggle the light

void setup() {
 pinMode(ledPin, OUTPUT); // declare the ledPin as as OUTPUT
 Serial.begin(9600);       // use the serial port
}

void loop() {
  // read the sensor and store it in the variable sensorReading:
  sensorReading = analogRead(knockSensor);

  // if the sensor reading is greater than the threshold:
  if (sensorReading >= threshold) {
    // toggle the status of the ledPin:
    ledState = !ledState;
    // update the LED pin itself:
    digitalWrite(ledPin, ledState);
    // send the string "Knock!" back to the computer,
    // followed by newline
```

```
        Serial.println("Knock!");
      }
      delay(100);  // delay to avoid overloading the serial port buffer
    }
```

Assuming all goes well, you will see how the piezo is detected when the Arduino Knock sketch toggles the onboard LED on or off.

The text `Knock!` is sent over the serial link every time the LED is toggled. Let's prove that this is being received correctly:

1. Launch the Serial Monitor by clicking Tools → Serial Monitor.
2. Check that the selected speed in the bottom-right corner of the Serial Monitor reads 9600 baud. If it does not read 9600 baud, change it using the drop-down list.
3. If the circuit is wired correctly, you should find that tapping hard on the piezo element causes the word `Knock!` to appear within the Serial Monitor window for every toggle of the LED.

If you don't get a `Knock!`, you might need to tap a bit harder. Another option is to look in the code for the following line, and change the sensitivity by modifying the threshold voltage at which the `Knock!` is triggered:

```
const int threshold = 100;  // threshold value to decide when the
                            // detected sound is a knock or not
```

Displaying Knock Messages in the Python Shell

Now, instead of watching the Knock messages in the Serial Monitor, you are going to run some code on the Raspberry Pi to read them, and decide what to do in response.

We'll be using the Python language for this job. Start by accessing the UART CP2102 serial device interactively to understand how Python commands work:

1. Close the Serial Monitor in the Arduino IDE to release the serial device. It may be necessary to close the Arduino IDE altogether to be sure the serial device is properly released. Launch LXTerminal in Raspbian.

2. Type the word python and press Enter to launch the Python interpreter. By running the interpreter without providing a filename, you have launched the Python shell.
3. Ask Python to import the Serial class from the serial library, which allows you to make connections to serial devices:

```
from serial import Serial
```

4. If successful, you should see no errors, and the cursor will return to the >>> prompt.
5. Now you can use the Serial class to connect to the UART CP2102 device. You may recall the path */dev/ttyUSB0*, which identified the serial device in the Arduino IDE. Let's use the same serial port to connect to the wearable using Python, passing it into the Serial class as a text string enclosed in single quotes:

```
connection = Serial('/dev/ttyUSB0')
```

6. If there were no errors, you now have a connection to the serial device, and you can read individual lines from it. The following Python line reads the next line from the serial device, then shows it in the shell window. If there's an unread Knock! line still waiting in the serial device from a previous tap, it should be shown. If there isn't a line to read, it will wait until the next line is sent to appear:

```
print(connection.readline())
```

7. You can now make the circuit send a new Knock! line by hitting the piezo.
8. Reload this command and execute it again (you can use the up arrow on your keyboard).
9. To make the behavior more responsive, let's create a block containing one step that will run inside a while loop forever by typing the following. You need to indent the second line using a Tab key, and finally leave a blank line by hitting Enter twice at the end:

```
while True:
    print(connection.readline())
```

10. After you enter the last blank line, you are completing the `while` loop. The cursor will not reappear, since the shell is busy running your code.
11. If all is well, `Knock!` should appear every time you tap the piezo.
12. When you want to stop the loop and get back to the shell, press Ctrl-C.

Loading a Python Script to Display Knock! Messages

Assuming you have unpacked our example source code directory to the desktop, you should be able to reproduce the behavior we just demonstrated in the Python shell, by loading all of the commands from a single file at once. When you are happy that taps are indeed triggering `Knock!` messages correctly, you can press Ctrl-C again to stop the script from running:

```
cd ~/Desktop/maker/Picussion/01_knock_serial/pi
python wired_serial.py
```

The code in that file looks like this:

```
from serial import Serial #  ❶

#  ❷
def sensor_loop(connection):
    while True: #  ❸
        line = connection.readline() #  ❹
        line = line.strip() #  ❺
        print line          #  ❻

wiredserial = Serial('/dev/ttyUSB0') #  ❼
sensor_loop(wiredserial)             #  ❽
```

❶ Loads the Python library for reading serial devices.

❷ Defines a function block called `sensor_loop`, which is invoked by passing it a serial device, and executes a sequence of steps using that device.

❸ Repeat while `True` is true, in other words, forever…

❹ Read bytes until newline, store in `line`.

❺ Remove leading, trailing whitespace from `line`.

❻ Makes text of the line appear on the console.

❼ Configure connection called `wiredserial`.

❽ Run steps in `sensor_loop` using `wiredserial`.

This code remains within the `while True:` block in `sensor_loop(...)` indefinitely, as prefigured by the comment about forever. In practice it keeps reading serial lines over and over again, and control doesn't return to the terminal console until there is an error, such as a keyboard interrupt or a serial disconnection.

Now, let's get the Bluetooth serial UART added to the wearable.

Exercise 3: Sending Knock Messages Wirelessly

It is time to add the HC-06 Bluetooth UART module to our wearable. Having demonstrated a piezo sensor and serial communication, you can combine these with a Bluetooth wireless serial communication to create a battery-powered, wearable tap sensor.

At this point in our project, there should be just one HC-06 wireless UART powered up within wireless range, using the factory configuration. Later in the chapter, you will be connecting to multiple devices in parallel, monitoring their sensors, and controlling their outputs all at the same time.

You don't need to modify the code running on the wearable at all. You will continue to run the unmodified Knock example sketch from the Arduino IDE on the wearables, and focus on the Python code running on the Pi.

Using Bluetooth for Programming

We are not using the wireless Bluetooth UART for programming the ATmega. It can, however, easily substitute for the wired UART after you have finished programming it.

You will be working with your new blinking wearable plus piezo and adding on a Bluetooth module (JY MCU) in this section (Figure 2-10). By the end of the exercise you will have a wireless wearable talking to your Raspberry Pi.

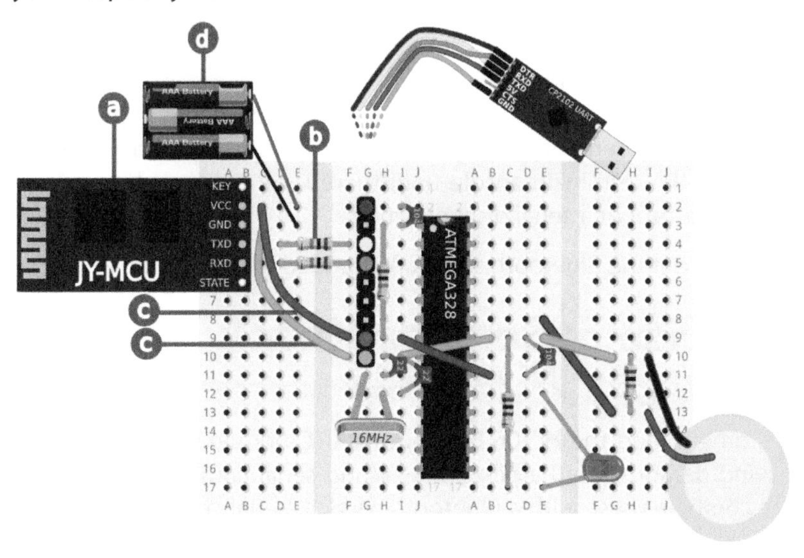

Figure 2-10. *Adding the Bluetooth module*

(a) Bluetooth UART module

The Bluetooth module should be presoldered with right-angle header pins. These pins are labeled in the sequence VCC, GND, TX, RX. You can slide the module directly into the solderless breadboard being sure to insert the labeled pins at the locations described in Table 2-9. The module will be pointing straight up from the board, although in Figure 2-10 the module has been drawn horizontally so you can better identify it.

Table 2-9. *Working on the left breadboard*

Component	From	To
(a) **Bluetooth UART module**	VCC (Power), B:2	RXD (Receive), B:5
(b) **10k ohm resistor**	TXD, D:4	TXD, F:4
(c) **Green wire**	ATmega GND, F:10	Bluetooth GND, C:3

Component	From	To
(c) **Red wire**	ATmega POW (Power), F:9	Bluetooth GND, C:2
(d) **Battery pack**	Black wire (Ground), E:3	Red wire (Power), E: 2

(b) 10k ohm resistor

The 3.3V pins are wired to a 5V signal, but we use the 10k ohm resistor to limit the electrical flow.

The resistors are only needed on the HC-06 receive line, but we use the resistor on both the transmit and receive lines in case you happen to wire TX and RX backwards. This prevents you from blowing up the module by mistake.

(c) Green wire and red wire

The wire color is not important, although if you use contrasting colors, it can help to avoid miswiring. Throughout this project we use red wires to indicate 5V (known as VCC) and green wires to indicate 0V (known as GND) to match the color of the ribbon cables provided with the CP2102 module.

(d) 3 x AAA battery pack

You have different options for power: a 3 x AAA battery pack (4.5V), the power lines connected from the CP2102 (5V), or a LiPo battery (3.7V), also known as a lithium polymer battery. In this project, we use the 3 x AAA battery pack when prototyping the breadboard version, but feel free to use LiPo batteries (for example, those provided with remote-control helicopters) because they are cheap, light, and powerful.

Make sure you don't have the CP2102 supply attached to the circuit at the same time as a battery pack.

Transmitting Wireless Messages

The Raspberry Pi doesn't have any built-in Bluetooth capabilities, so you'll have to attach a USB Bluetooth dongle (see the materials list at the beginning of the chapter) so that it can communicate wirelessly with the wearables.

Many HC-06 UARTs are fitted with an indicator LED, which blinks slowly to indicate that it has power and is discoverable. When the connection completes successfully, the light will remain on without blinking.

To pair your Pi with the HC-06 UART, perform the following in Raspbian on the Pi:

1. Launch a graphical Bluetooth configuration application called Blueman. Blueman is available under the LXDE start menu in the Preferences category. Alternatively, press Alt-F2, type **`blueman-manager`** in the box that appears, and press the Enter key.
2. Click the search button and wait for a module called `linvor` to appear in the list.
3. Right-click this entry and choose Pair.
4. When the dialog appears asking for a number as a PIN, type in **1234**. This is the default security code for connecting to the modules in their factory configuration.
5. Run the following in LXTerminal to launch the Python routine. The *bluetooth_serial.py* code should connect to the HC-06 UART, monitoring for new taps on the piezo, and printing `Knock!` to the screen when one is detected:

```
cd ~/Desktop/maker/Picussion/02_knock_wireless/pi
python bluetooth_serial.py
```

The following code from *bluetooth_serial.py* connects to the first bluetooth serial device with the name `linvor`, then prints out lines of text as they are received over the serial link:

```
from bluetooth import *

def reader_loop(reader):
    while True:
        line = reader.readline().strip()
        print line

# Look for addresses with these device names
serialnames = ['linvor','HC-06', 'Slinky']

# Detect bluetooth devices, filtering by name
for address in discover_devices():          # examine each device

    # check its name against the list
    if lookup_name(address) in serialnames:

        # Guess port, (saves service lookup)
        port = 1
        config = (address, port)            # store the address and port
        sock = BluetoothSocket()
        sock.connect(config)
        bluereader = sock.makefile("r")
        reader_loop(bluereader)

# If sensor_loop doesn't block, report problem
print "No Serial devices named " + ",".join(serialnames)
```

`Knock!` should have printed on your screen after running the script and tapping the piezo on your wearable. If you have multiple devices powered up in their factory configuration, it might be hard to know which one you've connected to because they are all named `linvor`. Use the LED toggle behavior described in Exercise 2 to verify that the circuit itself is detecting the taps and sending the serial.

Triggering Audio

Now that the Pi is successfully receiving Knock events wirelessly, let's make it do something useful with them. First of all, check that your Pi can produce sound:

1. Run the following from the LXTerminal, which should trigger a built-in recording for speaker testing: a woman's voice reading the words "Front, Center."

   ```
   aplay /usr/share/sounds/alsa/Front_Center.wav
   ```

2. If you heard her voice, proceed to the next step. If not, proceed to the following troubleshooting section.
3. Demonstrate that the sample is triggered when you tap the piezo element using the following commands in the LXTerminal window:

```
cd ~/Desktop/maker/Picussion/03_knock_audio_solo/pi
python solo_bluetooth.py
```

4. The slowly blinking light of the Bluetooth module should go solid when the connection has been made successfully. When taps are triggering sounds, press Ctrl-C to stop the script from running.

Troubleshooting Audio from the Pi

If you were unable to trigger audio on your Pi, then you'll want to read this section carefully. First, check that your audio output is configured correctly. Is there anything plugged into the 3.5mm audio jack? You can plug headphones or external speakers into the blue socket located on the side of the Pi.

If you have an HDMI screen attached to your Pi, it's possible that the audio is being routed to your screen instead. Is there a 3.5mm headphone socket on the screen itself, or perhaps a volume control that makes the speakers on your screen audible?

If you need to use an HDMI screen, but can't make the HDMI output audible, it's fairly straightforward to override the automatic routing mechanism for audio to force the use of the built-in 3.5mm jack. You can select audio outputs in the latest version of the `raspi-config` console setup utility. To be sure you're running the latest version you can run the following lines in LXTerminal:

```
sudo apt-get update
sudo apt-get upgrade raspi-config
sudo raspi-config
```

The first two lines upgrade the package (which relies on you having a network connection), and the third line runs the configuration utility. Choose whether the audio selection is automatic, force 3.5mm, or HDMI output under Advanced Options → Audio.

On the other hand, if you are receiving the error

```
main:682: audio open error: No such file or directory
```

you need to run the modprobe command and play the *Front_Center.wav* file, to make sure sound comes out:

```
sudo modprobe snd_bcm2835
aplay /usr/share/sounds/alsa/Front_Center.wav
```

You might also try running the `alsamixer` command to adjust your volume.

You should now have your audio working. It's time to configure your audio for the multiple wearables, which is crucial for the Picussion project.

Exercise 4: Talking Wearables

Before proceeding, you should repeat Exercises 1–3 and create a second wearable. Once you have two wearables triggering sound, it is time to get multiple sensors triggering multiple sounds for Picussion. To do this we will install the *multi_bluetooth.py* script onto our Raspberry Pi.

The following code specifies a whole list of paths to audio files, which are loaded into a list of sound samples. This version spawns a new thread to wait for text from the piezo on your wearable, and then goes to search for more sensors:

```
from threading import Thread

from pygame import mixer
mixer.init(frequency=22050, size=-16, channels=2, buffer=512)
# config to minimize delays

from bluetooth import *

soundpaths = [
    '/usr/share/sounds/alsa/Front_Center.wav',
    '/usr/share/sounds/alsa/Front_Left.wav',
    '/usr/share/sounds/alsa/Front_Right.wav',
    '/usr/share/sounds/alsa/Rear_Center.wav',
    '/usr/share/sounds/alsa/Rear_Left.wav',
    '/usr/share/sounds/alsa/Rear_Right.wav'
]

sounds = [pygame.mixer.Sound(path) for path in soundpaths]
```

```
def sensor_loop(connection, index):
    soundindex = index % len(sounds)
    sound = sounds[soundindex]
    while True:
        line = connection.readline().strip()
        if line.startswith("K"):    #look for knocks
            sound.play()

# Look for addresses with these device names
serialnames = ['linvor','HC-06', 'other']

# Detect bluetooth devices, filtering by name
addresses = [
    address for address in discover_devices()
    if lookup_name(address) in serialnames
]

# Set up multiple serial streams, managed by their own thread
if len(addresses) > 0:
    for (index, address) in enumerate(addresses):
        port = 1                    # Guess the port (this is a workaround
        config = (address, port)    # for unreadable RFCOMM service record)
        sock = BluetoothSocket()    # set up bluetooth serial
        sock.connect(config)
        blueserial = sock.makefile()
        thread = Thread(            # spawn new sensor thread
            target=sensor_loop,
            args=(serial, index)
        )
        thread.start()

# If no serial devices detected, report problem
else:
    print "No Serial devices named " + ",".join(serialnames)
```

Using Python Threading for Multiple Wearables

You've been using `readline()` in the code examples to get a line of text from a serial connection. You may have noticed that `readline()` is configured as a blocking function. That is to say, the execution of commands stops until a new line of text can be read from the connection. Given that you can only run one command at a time with `readline()`, this creates the problem that while you're waiting for a line of text from one sensor, you are ignoring another sensor. Instead, you need to create a responsive sound-triggering system that can monitor all of the sensors and respond to whichever one is tapped.

To solve this problem, we'll use the Python `threading` module (see the previous *multi_bluetooth.py* code). This module launches more than one thread of execution, allowing you to execute multiple commands at the same time.

Running Audio on Multiple Wearables

Let's start by powering up our wearables by switching on the AAA battery pack that powers each wearable, which will make the HC-06 UARTs blink. Launch an LXTerminal window and type the following lines, each followed by the Enter key:

```
cd ~/Desktop/maker/Picussion/04_knock_audio_multi/pi
python multi_bluetooth.py
```

Once you launch this Python script (as the devices are detected and connections are made), you should see individual lights on the HC-06 UARTs go from blinking to steady on. After each light stops blinking, tap the piezo for that wearable and an audio sample will play through your Pi's audio output.

Exercise 5: One Man Band

We will be rounding off this chapter by adding on an ADXL345 accelerometer breakout board to each wearable, allowing orientation to be sensed.

Adding an Accelerometer

Follow both Figure 2-11 and Table 2-10 to add the accelerometer.

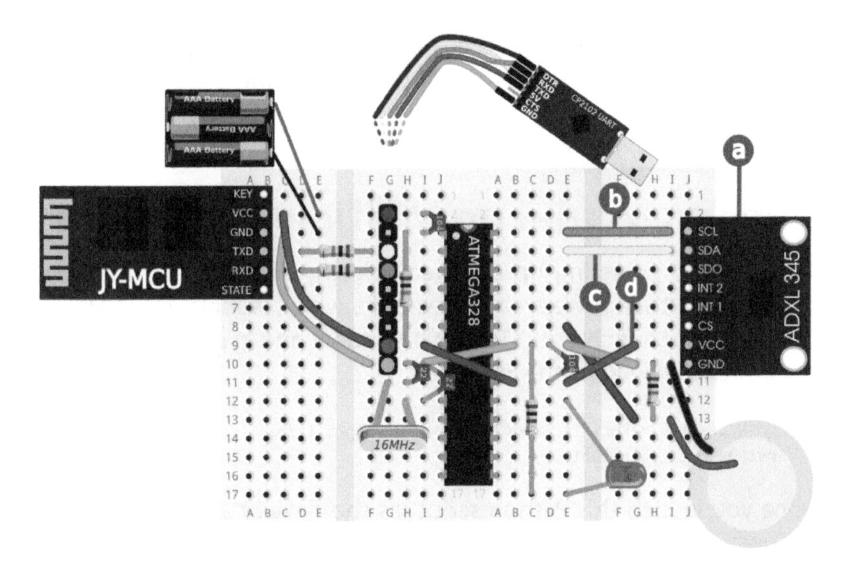

Figure 2-11. *Adding the accelerometer*

Table 2-10. *Right breadboard*

Component	From	To
(a) **Accelerometer (ADXL 345)**	SCL (Serial Communications Clock), J:3	SDA (Serial Data), J:10
(b) **I2C orange wire**	ATmega AI5 (Analog Input Pin 5), E:3	Accelerometer SCL, I:3
(c) **I2C yellow wire**	ATmega AI4 (Analog Input Pin 4), E:4	Accelerometer SDA, I:4
(d) **Red wire**	ATmega POW (Power), E:11	Accelerometer VCC (Power), G:9

(a) Accelerometer (ADXL 345)

The ADXL345 reports three numbers, one per axis. These numbers indicate how much weight the ADXL345 is experiencing in a given direction. If it is upside down, it will report negative weight. You will learn how to attach your ADXL345, and how to modify the code on the wearable to be able to send the x, y, and z values of its three-axis orientation over Bluetooth.

(b), (c) Orange and yellow wires

The ATmega chip has support for Inter-Integrated Circuit (I2C) communication. This standard allows it to be attached to a chain of other integrated circuit devices (ICs), sending and receiving structured digital messages, with every IC connected to the same two signal wires, colored orange and yellow in our circuit, that are known as Data and Clock. The Accelerometer IC we are using, known as an ADXL345, is an accelerometer that can use the I2C protocol to send information about what it's sensing.

(d) Red wire

The ADXL345 needs a power supply as well as signaling wires. The connection to ground is already in place from the piezo build earlier, so this red wire completes the power supply.

Detecting Orientation

You need to update the behavior of the wearable so that it knows to read accelerometer data and send it out over the Bluetooth serial link. Once the wearable is sending the extra data, we can build a behavior in Python on the Pi. This behavior triggers audio depending upon the orientation of the sensor. Here is the *gravity_trigger.py* script that achieves this:

```
import math
import numpy

from threading import Thread

from pygame import mixer
mixer.init(frequency=22050, size=-16, channels=2, buffer=512)
# minimizes delays

from bluetooth import *

# class to trigger sounds based on angle between current vector
# & target vector
class Hotspot:

    def __init__(self, vector, path):
        self.vector = vector
        self.path = path
        self.matrix = numpy.array(vector) # turn into matrix for math
        self.sound = mixer.Sound(path) # turn filepaths into
                                       # playable samples
```

```
        self.active = False

    # vector dot_product divided by product of
    # magnitude=cosine of angle between
    def get_angle(self, othermatrix):
        return math.degrees(
            math.acos(
                self.matrix.dot(othermatrix) / (
                    numpy.sqrt(self.matrix.dot(self.matrix)) *
                    numpy.sqrt(othermatrix.dot(othermatrix))
                )
            )
        )

    # refresh the angle with a new update of accelerometer data
    def update(self, sensor_matrix):
        self.angle = self.get_angle(sensor_matrix)
        if self.is_hot():
            # hotspot should be active
            if not(self.active):
                print "Activate:", self.vector
                self.sound.play()
                self.active = True
        else:
            # hotspot should not be active
            if self.active:
                print "Deactivate  :", self.vector
                self.sound.stop()
                self.active = False

    def is_hot(self):
        return self.angle < 30

spots = [
    Hotspot([0,0,1],  '/usr/share/sounds/alsa/Front_Center.wav' ),
    Hotspot([0,1,0],  '/usr/share/sounds/alsa/Front_Left.wav'   ),
    Hotspot([1,0,0],  '/usr/share/sounds/alsa/Front_Right.wav'  ),
    Hotspot([0,0,-1], '/usr/share/sounds/alsa/Rear_Center.wav'  ),
    Hotspot([0,-1,0], '/usr/share/sounds/alsa/Rear_Left.wav'    ),
    Hotspot([-1,0,0], '/usr/share/sounds/alsa/Rear_Right.wav'   )
]

# Look for addresses with these device names
serialnames = ['linvor','HC-06', 'Slinky']

# List bluetooth devices, filtering by name
def list_available_addresses():
    return [
        address for address in discover_devices()
        if lookup_name(address) in serialnames
    ]
```

```
def connect_link(address, modes=("r","w")):
    print("Connecting to " + address)
    port = 1    # Guess port (workaround unreadable service record)
    config = (address, port)
    sock = BluetoothSocket()
    sock.connect(config)
    return [sock.makefile(mode,0) for mode in modes]

addresses = list_available_addresses()

# Set up multiple serial streams, managed by their own thread
if len(addresses) > 0:

    # connect to first sensor
    (reader,writer) = connect_link(addresses[0])


    while True:
        matrix = None
        while matrix == None:
            line = reader.readline()
            if(line[0:2]=="A:"):                # detect prefix
                line = line[2:]                 # remove prefix
                vals = line.split(',')          # separate text at commas

                # turn text into numbers
                vals = [float(val) for val in vals]

                matrix = numpy.array(vals)      # turn numbers into matrix
                for spot in spots:
                    spot.update(matrix)

# If no serial devices detected, report problem
else:
    print "No Serial devices named " + ",".join(serialnames)
```

Adding the KnockLedControl Sketch

We will be integrating Adafruit's Unified Sensor Driver for the ADXL345 with the KnockLedControl sketch.

First, here's how you add the KnockLedControl sketch:

1. Assuming you have unpacked the example source code directory on the Raspberry Pi desktop, you can open the modified Knock sketch from */home/pi/Desktop/maker/Picussion/05_knock_leds/shrimp/KnockLedControl* as shown here:

    ```
    /* Knock Sensor plus LED control
    ```

```
  This sketch reads a piezo element to detect a knocking sound.
  It reads an analog pin and compares the result to a set threshold.
  If the result is greater than the threshold, it writes
  "Knock!" to the serial port.

  The sketch responds to a Serial message "on" followed by a newline
  by turning on its LED, and responds to "off" followed by a newline
  by turning off its LED.

  Based on http://www.arduino.cc/en/Tutorial/Knock created
  25 Mar 2007 by David Cuartielles <http://www.0j0.org>

  modified 30 Aug 2011 by Tom Igoe
  modified 25 May 2014 by Cefn Hoile

  This example code is in the public domain.

 */

// these constants won't change:
const int ledPin = 13;       // led connected to digital pin 13
const int knockSensor = A0; // the piezo is connected to analog pin 0
const int threshold = 50;    // threshold value to decide
                             // when the detected
                             // sound is a knock or not

unsigned long lastKnock = 0; // the last time in milliseconds
                             // that a knock was detected
const unsigned long ignoreKnock = 50; // how long to ignore knocks
                                      // after one is
                                      // detected (debounce)

// these variables will change:
int sensorReading = 0;       // variable to store the value read
                             // from the sensor pin
int ledState = LOW;          // variable used to store the last LED
                             // status, to toggle the light

void setup() {
 pinMode(ledPin, OUTPUT);  // declare the ledPin as as OUTPUT
 Serial.begin(9600);       // use the serial port
}

void loop() {
  // read the sensor and store it in the variable sensorReading:
  sensorReading = analogRead(knockSensor);

  // if the sensor reading is greater than the threshold:
  if ((sensorReading >= threshold) &&
      ((millis() - lastKnock) > ignoreKnock))
  {
    lastKnock = millis();
```

```
    // send the string "Knock!" back to the computer,
    // followed by newline
    Serial.println("Knock!");
  }

  while(Serial.available()){
    String serialString = Serial.readStringUntil('\n');

    if(serialString.equals("on")){
      ledState = HIGH;
    }
    else if(serialString.equals("off")){
      ledState = LOW;
    }
    else{
      Serial.print("Unexpected text:");
      Serial.print(serialString);
      Serial.println();
    }
  }

  digitalWrite(ledPin, ledState);

}
```

2. Attach the CP2102 (the USB to serial module) to each of your wearables in turn, and upload this KnockLedControl sketch.
3. After successfully uploading the modified Knock sketch, you can try turning on and off the LED by opening the Arduino IDE Serial Monitor.
4. Be sure to select newline and 9600 baud from the drop-down menus at the bottom of the window.
5. Typing the word **on**, followed by the Enter key, should make the LED go on. Typing the word **off**, followed by the Enter key, should make it go off again.

Now it's time to add the Adafruit driver. We will first navigate to the libraries folder of your Arduino sketchbook, and then download the driver using a code versioning tool called git. To install git, if it isn't already installed, type the following line in LXTerminal, and press Enter:

```
sudo apt-get update; sudo apt-get install git
```

If you don't know where your Arduino sketchbook is located, you can click File → Preferences to see the full directory path. The default path on a Raspbian install is */home/pi/sketchbook*, so we'll go there, create the needed directories, and download the libraries. Type the following lines in LXTerminal, pressing Enter after each one and waiting for it to complete:

```
cd /home/pi/sketchbook
mkdir -p libraries
cd libraries
git clone https://github.com/adafruit/Adafruit_Sensor.git Adafruit_Sensor
git clone https://github.com/adafruit/Adafruit_ADXL345.git
Adafruit_ADXL345_U
```

After running these commands, do the following:

1. Restart the Arduino IDE. You should find a new example sketch called sensortest under File → Examples → Adafruit_ADXL345_U.
2. Upload this sensortest sketch to one of your wearables with an ADXL345 onboard and observe the serial data coming back over the CP2102 UART.
3. Be sure to select 9600 baud as the speed in the Arduino Serial Monitor.

You should see information printing out about the sensor's configuration, followed by updates of the X, Y, and Z axis weight readings every half a second.

The final KnockLedAccel sketch is based on merging our KnockLedControl sketch with the sensortest sketch, disposing of the extra behaviors from the Adafruit sketch that we don't need (like reporting the sensor configuration). Here is the contents of the sketch:

```
/* Knock Sensor plus LED control plus ADXL345 accelerometer
   This sketch reads a piezo element to detect a knocking sound.
   It reads an analog pin and compares the result to a set
   threshold.

   If the result is greater than the threshold, it writes
   "K:" to the serial port, followed by the analog reading.
   The sketch responds to a Serial message "on" followed by a
   newline by turning on its LED, and responds to "off" followed
   by a newline by turning off its LED.

   Additionally it periodically reads an ADXL345 digital
   accelerometer over I2C, reading X,Y,Z values, then writing
   "A:" to the serial port out, followed by the three axis readings.
```

```
   Based on http://www.arduino.cc/en/Tutorial/Knock created
   25 Mar 2007 by David Cuartielles <http://www.0j0.org>

   modified 30 Aug 2011 by Tom Igoe
   modified 25 May 2014 by Cefn Hoile
   This example code is in the public domain.
*/

#include <Wire.h>
#include <Adafruit_Sensor.h>
#include <Adafruit_ADXL345_U.h>

/* Assign a unique ID to this sensor at the same time */
Adafruit_ADXL345_Unified accel = Adafruit_ADXL345_Unified(12345);

// these constants won't change:
const int ledPin = 13; // led connected to digital pin 13
const int knockSensor = A0; // the piezo is connected to
                            // analog pin 0
const int threshold = 50; // threshold value to decide
                          // when the detected
                          // sound is a knock or not

unsigned long lastKnock = 0; // the last time in milliseconds
                             // that a knock was detected
const unsigned long ignoreKnock = 50; // how long to ignore knocks
                                      // after one is detected
                                      // (debounce)

// these variables will change:
int sensorReading = 0; // variable to store the value read
                       // from the sensor pin
int ledState = LOW; // variable used to store the last LED
                    // status, to toggle the light
void setup() {
   pinMode(ledPin, OUTPUT); // declare the ledPin as as OUTPUT
   Serial.begin(9600); // use the serial port

  /* Initialise the sensor */
  if(!accel.begin())
  {
    /* There was a problem detecting the ADXL345
       ... check your connections */
    Serial.println(
      "Ooops, no ADXL345 detected ... Check your wiring!");
    while(1);
  }
  accel.setRange(ADXL345_RANGE_16_G);
}
void loop() {

  /* Display the acceleration (in m/s^2) */
```

```
  sensors_event_t event;
  accel.getEvent(&event);
  Serial.print("A: ");
  Serial.print(event.acceleration.x); Serial.print(",");
  Serial.print(event.acceleration.y); Serial.print(",");
  Serial.println(event.acceleration.z);

  // read the sensor and store it in the variable sensorReading:
  sensorReading = analogRead(knockSensor);

  // if the sensor reading is greater than the threshold:
  if ((sensorReading >= threshold)
      && ((millis() - lastKnock) > ignoreKnock)) {
    lastKnock = millis();
    // send the string "Knock!" back to the computer,
    // followed by newline
    Serial.print("K:");
    Serial.println(sensorReading);
  }
  while(Serial.available()){
    String serialString = Seri! al.readStringUntil('\n');

    if(serialString.equals("on")){
      ledState = HIGH;
    }
    else if(serialString.equals("off")){
      ledState = LOW;
    }
    else{
      Serial.print("Unexpected text:" );
      Serial.print(serialString);
      Serial.println();
    }
  }
  digitalWrite(ledPin, ledState);

}
```

Let's briefly walk through some of the changes that have been made. The first change to KnockLedControl is to import the Wire library, which gives you general I2C support, and the Adafruit libraries, to specifically support the accelerometer. The #include directive asks the compiler to put the whole of the Wire and Adafruit files in the top of your code, as if you had written the code there yourself, but without the extra code cluttering up the program:

```
#include <Wire.h>
#include <Adafruit_Sensor.h>
#include <Adafruit_ADXL345_U.h>
```

You normally don't need to look at the code in those files directly, but you will be using some of the functions defined inside them, such as this function call, which creates a new object named `accel` to represent your accelerometer. With this object you can later configure the sensitivity of the accelerometer, and retrieve data from it:

```
Adafruit_ADXL345_Unified accel = Adafruit_ADXL345_Unified(12345);
```

We copied the initialization code directly from the `setup()` function in sensortest into our own `setup()` function:

```
/* Initialise the sensor */
if(!accel.begin())
{
  /* There was a problem detecting the ADXL345
     ... check your connections */
  Serial.println("Ooops, no ADXL345 detected ... Check your wiring!");
  while(1);
}

accel.setRange(ADXL345_RANGE_16_G);
```

Finally, in the main loop, the code will regularly ask the `accel` object to populate a `sensors_event` data structure for you, which will include updated x, y, and z values. We use `Serial.print()` to send these over the Bluetooth serial link, prefixed by "A:", with commas separating the numbers and a `println(...)` to finish the record with a newline character:

```
sensors_event_t event;
accel.getEvent(&event);
Serial.print("A: ");
Serial.print(event.acceleration.x); Serial.print(",");
Serial.print(event.acceleration.y); Serial.print(",");
Serial.println(event.acceleration.z);
```

The sketch continues to report knocks and responds to LED on and off commands. For consistency, we're outputting the piezo readings as shown here, where the number indicates the intensity of the voltage spike detected, ranging between 0 and 1023:

```
K:150
```

Now, let's move on to creating our hotspots that will allow Picussion gestures to trigger selected audio files, such as having your hands out to mimic the playing of a trumpet.

Processing Accelerometer Data on the Pi

This example is a simple demonstration of triggering audio according to the orientation of a wearable:

```
import math
import numpy

from threading import Thread

from pygame import mixer
mixer.init(frequency=22050, size=-16, channels=2, buffer=512)
# minimizes delays

from bluetooth import *

# class to trigger sounds based on angle between
# current vector & target vector
class Hotspot:

    def __init__(self, vector, path):
        self.vector = vector
        self.path = path
        self.matrix = numpy.array(vector) # turn into matrix for math
        self.sound = mixer.Sound(path) # turn filepaths into
                                       # playable samples
        self.active = False

    # vector dot_product divided by product of
    # magnitude=cosine of angle between
    def get_angle(self, othermatrix):
        return math.degrees(
            math.acos(
                self.matrix.dot(othermatrix) / (
                    numpy.sqrt(self.matrix.dot(self.matrix)) *
                    numpy.sqrt(othermatrix.dot(othermatrix))
                )
            )
        )

    # refresh the angle with a new update of accelerometer data
    def update(self, sensor_matrix):
        self.angle = self.get_angle(sensor_matrix)
        if self.is_hot():
            # hotspot should be active
            if not(self.active):
                print "Activate:", self.vector
                self.sound.play()
                self.active = True
```

```
        else:
            # hotspot should not be active
            if self.active:
                print "Deactivate  :", self.vector
                self.sound.stop()
                self.active = False

    def is_hot(self):
        return self.angle < 30

spots = [
    Hotspot([0,0,1],  '/usr/share/sounds/alsa/Front_Center.wav' ),
    Hotspot([0,1,0],  '/usr/share/sounds/alsa/Front_Left.wav'   ),
    Hotspot([1,0,0],  '/usr/share/sounds/alsa/Front_Right.wav'  ),
    Hotspot([0,0,-1], '/usr/share/sounds/alsa/Rear_Center.wav'  ),
    Hotspot([0,-1,0], '/usr/share/sounds/alsa/Rear_Left.wav'    ),
    Hotspot([-1,0,0], '/usr/share/sounds/alsa/Rear_Right.wav'   )
]

# Look for addresses with these device names
serialnames = ['linvor','HC-06', 'Slinky']

# List bluetooth devices, filtering by name
def list_available_addresses():
    return [
        address for address in discover_devices()
        if lookup_name(address) in serialnames
    ]

def connect_link(address, modes=("r","w")):
    print("Connecting to " + address)
    port = 1    # Guess port (workaround unreadable service record)
    config = (address, port)
    sock = BluetoothSocket()
    sock.connect(config)
    return [sock.makefile(mode,0) for mode in modes]

addresses = list_available_addresses()

# Set up multiple serial streams, managed by their own thread
if len(addresses) > 0:

    # connect to first sensor
    (reader,writer) = connect_link(addresses[0])


    while True:
        matrix = None
        while matrix == None:
            line = reader.readline()
            if(line[0:2]=="A:"):             # detect prefix
                line = line[2:]              # remove prefix
```

```
            vals = line.split(',')       # separate text at commas

            # turn text into numbers:
            vals = [float(val) for val in vals]

            matrix = numpy.array(vals) # turn numbers into matrix
            for spot in spots:
                spot.update(matrix)

# If no serial devices detected, report problem
else:
    print "No Serial devices named " + ",".join(serialnames)
```

In this *gravity_trigger.py* script, we have imports of `math` and `numpy`, which provide access to simple mathematical operations and matrix mathematics, respectively. The numbers we get from the accelerometer can be brought together as a three-dimensional vector or matrix. The matrix consists of three numbers that indicate the direction and strength of the g-force that the sensor is experiencing.

We test the angle of the current gravity vector (which way is down) relative to six reference gravity vectors, corresponding with the six faces of a cube. Each instance of the `Hotspot` class monitors how close the device is to a specific reference angle, and triggers audio as the device gets close to that angle.

Each hotspot is assigned a vector (its trigger orientation), a path (the path to the audio file that will be triggered), and a test criterion that will cause the audio sample to play when it is satisfied. Each hotspot is triggered when the angle of gravity is within 30 degrees of any one of these targets (Figure 2-12). For this reason, the default trigger criterion is defined as:

```
def default_test(spot):
return spot.angle < 30
```

The `Hotspot` class has a convenience method, called `getangle`, that accepts a `numpy` matrix and reports the angle relative to its trigger vector:

```
def get_angle(self, othermatrix):
    return math.degrees(
        math.acos(
            self.matrix.dot(othermatrix) / (
                numpy.sqrt(self.matrix.dot(self.matrix)) *
                numpy.sqrt(othermatrix.dot(othermatrix))
```

```
            )
        )
    )
```

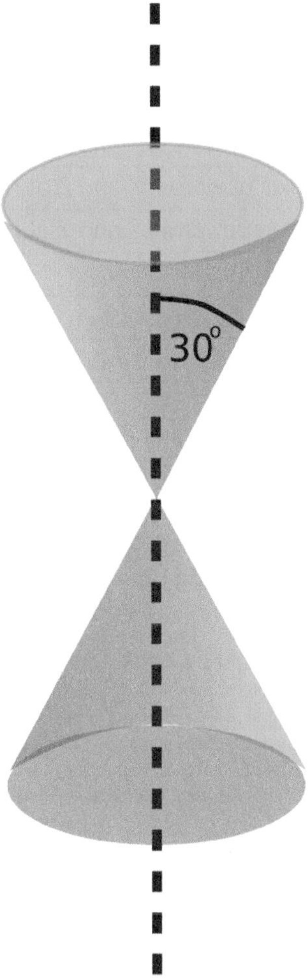

Figure 2-12. *A 30-degree trigger cone around target orientation*

In our main loop, we'll keep updating the angles of all of the hotspots as new accelerometer data comes in. As each hotspot gets an update, it can decide whether or not the current orientation should trigger its sample.

To create all of the hotspots, we need to figure out the orientation vectors for each face, and the sound samples to trigger them. Imagine that gravity was measured at 1g, and we turned the wearable to show each face turned upwards, named from A–F. The accelerometer readings we would get are:

Axes	X	Y	Z	Description
A	0	0	1	Z axis upwards, X and Y at right angles to gravity
B	0	1	0	Y axis upwards, X and Z at right angles to gravity
C	1	0	0	X axis upwards, Y and Z at right angles to gravity
D	0	0	–1	Like A, upside down
E	0	–1	0	Like B, upside down
F	–1	0	0	Like C, upside down

In the `spots` list, you can see a series of hotspots being explicitly created, which use six very convenient Alsa test sounds that are distributed as part of the Raspbian OS. These sounds are a female voice reading "Front Center," "Front Left," "Front Right," "Rear Center," "Rear Left," and "Rear Right," depending upon which side is triggered:

```
spots = [
    Hotspot([0,0,1],  '/usr/share/sounds/alsa/Front_Center.wav' ),
    Hotspot([0,1,0],  '/usr/share/sounds/alsa/Front_Left.wav'   ),
    Hotspot([1,0,0],  '/usr/share/sounds/alsa/Front_Right.wav'  ),
    Hotspot([0,0,-1], '/usr/share/sounds/alsa/Rear_Center.wav'  ),
    Hotspot([0,-1,0], '/usr/share/sounds/alsa/Rear_Left.wav'    ),
    Hotspot([-1,0,0], '/usr/share/sounds/alsa/Rear_Right.wav'   )
]
```

If you want to change the audio triggered by each side of the cube, it is easy to replace these file paths with the paths to your own samples.

Now we come to the main part of the program, which will search for HC-06 serial devices and connect to the first one it encounters. The program will read lines from serial, filtering for accelerometer data (lines beginning with A:), and trying to extract the x, y, and z numbers from each line. Once each number has been read, each hotspot with the new vector information is updated, leaving it to the hotspot to trigger if its criterion is met. Follow these steps:

1. Open the modified Knock sketch from */home/pi/Desktop/maker/Picussion/06_gravity_audio/shrimp/KnockLedAccel*.
2. Attach the CP2102 to each of your wearables in turn, and upload this sketch.
3. Now you can run the audio to prove everything is working by typing the following lines within an LXTerminal window, pressing the Enter key after each line:

```
cd ~/Desktop/maker/Picussion/06_gravity_audio/pi/
python gravity_trigger.py
```

As before, the slowly blinking light of the Bluetooth module should go solid on each of the wearables when the connection has been made successfully. Once the connection has completed, turning the sensor module should trigger playback of the six *.wav* files.

Running the One Man Band

The final Picussion application is the One Man Band. This combines the orientation sensing of multiple wearable sensor modules on different parts of the body at the same time. By making predefined gestures of certain instruments, you will then have *.wav* files play the recorded sound from that actual instrument.

To make the sounds work with each other seamlessly, we're going to start with a multitrack with six different tracks recording on a loop. We worked with a downloaded version of Marvin Gaye's "Heard it on the Grapevine" in which the different instrumental tracks have been digitally separated. We left out the lead vocals and the chorus so we could sing along.

This table describes the gestures we designed the application to detect, corresponding with each named track from the multitrack archive. Most of the gestures rely on two wearables mounted to the back of the wrist, but if a third wearable is available, it should be strapped to the top of the left thigh to replicate the classic one man band drum mechanism.

Instrument	Gesture
Organ, Electric Piano	Right wrist and left wrist, both palm downwards

Bass Guitar	Right wrist across chest, left wrist palm upwards
Rhythm Guitar	Right wrist across chest, left wrist palm backwards
Strings, French Horns	Left wrist and right wrist each with palm angled 45 degrees down
Tambourine, Congas	Left wrist across chest, palm backwards, right wrist downwards, palm backwards
Kit Drums	Right thigh at 45 degrees

How the One Man Band Works

This program scans for Bluetooth UART devices in the area until it has successfully connected to enough accelerometer sensors. Then it starts monitoring for audio-triggering gestures using those sensors.

The program will initially indicate that it's waiting for two sensors to become available; one attached to each of the performer's wrists. The user should turn the sensors on one by one, and wait for each to be detected.

If they are detected in a predictable order, it's easy for the Python program (and for you) to know which one is which. Once all of the sensors have been detected, the samples begin playing the multitrack recording at the same time, but with zero volume. This way each detected gesture will activate a single instrument at a time.

Here is the full program, *gesture_trigger.py*:

```
import math
import numpy

from threading import Thread

from pygame import mixer
mixer.init(frequency=22050, size=-16, channels=2, buffer=512)
# minimizes delays

from bluetooth import *
from time import sleep

num_links = 2
vectors = [None for link in range(0,num_links)]
```

```
class Synth:

    def __init__(self,path,*tests):
        self.path = path
        self.sound = mixer.Sound(path)
        self.tests = tests
        self.active = False

    def update(self, vectors):
        if None in vectors: # vectors are not yet all known
            return
        else:
            if False not in [test(*vectors) for test in self.tests]:
                if not self.active:
                    print "Activating " + self.path
                    self.active = True
                    self.set_volume(1)
            else:
                if self.active:
                    print "Deactivating " + self.path
                    self.active = False
                    self.set_volume(0)

    def set_volume(self, vol):
        self.sound.set_volume(vol)

# returns average of several vectors
def between(*vectors):
    return numpy.array([sum(col) / len(vectors) for col in zip(*vectors)])

# vector dot_product divided by product of magnitude=cosine of
# angle between
def get_angle(matrix, othermatrix):
    return math.degrees(
        math.acos(
            matrix.dot(othermatrix) / (
                numpy.sqrt(matrix.dot(matrix)) *
                numpy.sqrt(othermatrix.dot(othermatrix))
            )
        )
    )


faces = [
    [0,0,-1.0],
    [0,-1.0,0],
    [-1.0,0,0],
    [0,0,1.0],
    [0,1.0,0],
    [1.0,0,0]
]
```

```
# names correspond to sides of the housing (ADXL orientations)
[
    DOWNSIDE_DOWN,
    FRONTSIDE_DOWN,
    RIGHTSIDE_DOWN,
    UPSIDE_DOWN,
    BACKSIDE_DOWN,
    LEFTSIDE_DOWN
] = [numpy.array(face) for face in faces]

cone = 30

# Right wrist, Left wrist, both palm downwards
def test_keyboard(left_wrist, right_wrist, left_thigh=BACKSIDE_DOWN):
    return (
        get_angle(right_wrist, DOWNSIDE_DOWN) < cone
            and
        get_angle(left_wrist, DOWNSIDE_DOWN) < cone
    )

# Right wrist across chest, palm backward, Left wrist palm backward
def test_bass(left_wrist, right_wrist, left_thigh=BACKSIDE_DOWN):
    return (
        get_angle(right_wrist, RIGHTSIDE_DOWN) < cone
            and
        get_angle(left_wrist, FRONTSIDE_DOWN) < cone
    )

# Right wrist across chest palm backward, Left wrist palm upward
def test_guitar(left_wrist, right_wrist, left_thigh=BACKSIDE_DOWN):
    return (
        get_angle(right_wrist, RIGHTSIDE_DOWN) < cone
            and
        get_angle(left_wrist, UPSIDE_DOWN) < cone
    )

# Left wrist and right wrist each with palm angled 45 degrees down
def test_horns(left_wrist, right_wrist, left_thigh=BACKSIDE_DOWN):
    target_vector = between(FRONTSIDE_DOWN,DOWNSIDE_DOWN)
    return (
        get_angle(right_wrist, target_vector) < cone
            and
        get_angle(left_wrist, target_vector) < cone
    )

# Left wrist across chest, palm backwards, right wrist downward,
# palm backwards
def test_handdrum(left_wrist, right_wrist, left_thigh=BACKSIDE_DOWN):
    return (
        get_angle(right_wrist, BACKSIDE_DOWN) < cone
            and
```

```
        get_angle(left_wrist, LEFTSIDE_DOWN) < cone
    )

# Right thigh at 45 degrees
def test_kickdrum(left_wrist, right_wrist, left_thigh=BACKSIDE_DOWN):
    return (
        get_angle(left_thigh,BACKSIDE_DOWN) > cone
    )

# intro sequence is organ, drums (cymbals), tambourine, lead guitar
synths = [
    Synth("./tracks/Organ, Electric Piano.wav", test_keyboard),
    Synth("./tracks/Bass.wav", test_bass),
    Synth("./tracks/Guitar.wav", test_guitar),
    Synth("./tracks/Strings, French Horns.wav", test_horns),
    Synth("./tracks/Tambourine, Congas.wav", test_handdrum),
    Synth("./tracks/Drums.wav", test_kickdrum)
]

# stores sensed gravity vectors and updates program state
def update_vector(index, vector):
    vectors[index] = vector
    for synth in synths:
        synth.update(vectors)

link_addresses = []

# Look for addresses with these device names
serialnames = ['linvor','HC-06', 'Slinky']

# List bluetooth devices, filtering by name
def list_available_addresses():
    return [
        address for address in discover_devices()
        if address not in link_addresses and
        lookup_name(address) in serialnames
    ]

def connect_link(address, modes=("r","w")):
    print("Connecting to " + address)
    port = 1    # Guess port (workaround unreadable service record)
    config = (address, port)
    sock = BluetoothSocket()
    sock.connect(config)
    return [sock.makefile(mode,0) for mode in modes]

def reader_loop(reader, index):
    while True:
        line = reader.readline().strip()
        if(line[0:2]=="A:"):                    # detect prefix
            line = line[2:]                     # remove prefix
```

```
            # split text at commas:
            vals = [float(val) for val in line.split(',')]
            # turn numbers into matrix:
            update_vector(index, numpy.array(vals))

    # keep searching until there are enough to link to
    while True:
        print "Awaiting " + str(num_links - len(link_addresses)) + " links"
        fresh_addresses = list_available_addresses()
        link_addresses.extend(fresh_addresses)
        if len(link_addresses) < num_links :
            sleep(1)
        else:
            break

    # connect to the links
    links = [connect_link(address) for address in link_addresses[:num_links]]

    print "Connected to all required sensor units, now starting music"

    # spawn threads to handle sensor updates
    for (index, (reader, writer)) in enumerate(links):
        thread = Thread(                    # spawn new sensor thread
            target=reader_loop,
            args=(reader, index)
        )
        thread.daemon = True                # stop if main thread stops
        thread.start()

    # trigger sounds on a loop
    for synth in synths:
        synth.sound.play(-1)
        synth.sound.set_volume(0)

    # wait while sensor readings trigger synths on and off
    while True:
        sleep(0.05)
```

In this code we define a Synth class, which couples the playback of an audio file to a test function, and creates several instances of Synth—one per track for each of the audio files in our multitrack Marvin Gaye example. The tracks and gestures might be different for your own choice of music:

```
synths = [
    Synth("./tracks/Organ, Electric Piano.wav", test_keyboard),
    Synth("./tracks/Bass.wav", test_bass),
    Synth("./tracks/Guitar.wav", test_guitar),
    Synth("./tracks/Strings, French Horns.wav", test_horns),
    Synth("./tracks/Tambourine, Congas.wav", test_handdrum),
```

```
        Synth("./tracks/Drums.wav", test_kickdrum)
    ]
```

To define tests for gestures, we are using a set of reference vectors:

```
    faces = [
        [0,0,-1.0],
        [0,-1.0,0],
        [-1.0,0,0],
        [0,0,1.0],
        [0,1.0,0],
        [1.0,0,0]
    ]
```

We figured out how these reference vectors relate to orientations of the wearable by running the Knock Wireless example earlier, which just reports all of the sensor readings to the console. With the knowledge of which orientation is which, we then stored each of these orientation vectors within a named `numpy.array` so that we can calculate vector angles with them very quickly using `numpy` operations:

```
    [
        DOWNSIDE_DOWN,
        FRONTSIDE_DOWN,
        RIGHTSIDE_DOWN,
        UPSIDE_DOWN,
        BACKSIDE_DOWN,
        LEFTSIDE_DOWN
    ] = [numpy.array(face) for face in faces]
```

Front, back, left, right, up, and down are judged from the point of view of someone wearing a sensor on the back of their wrist holding their hand out flat, palm downwards. We chose names according to which face of the wearable sensor box is downwards, which will help us define gestures later.

The test functions inspect the current position of the wearables and detect gestures, as illustrated by the `test_guitar` function:

```
    def test_guitar(left_wrist, right_wrist, left_thigh=BACKSIDE_DOWN):
        return (
            get_angle(right_wrist, RIGHTSIDE_DOWN) < cone
                and
            get_angle(left_wrist, UPSIDE_DOWN) < cone
        )
```

This function accepts a series of `numpy.array` vectors describing the orientation of each limb's wearable sensor. It then tests if the angle between its current orientation and the target vector for that limb is less than a trigger value, which defines a trigger cone of angles that would satisfy the gesture. If all of the limb positions match with the gesture, the synth is triggered. By default, the cone angle used is 30 degrees across.

Proper Gestures for the One Man Band

Let's now run our One Man Band program. Once it's running, you can follow along with the gestures in Figures 2–13 through 2–18. Type the following commands in an LXTerminal window, pressing Enter after each one. As always, press Ctrl-C again to stop the script from running:

```
cd ~/Desktop/maker/Picussion/07_gesture_audio/pi
python gesture_trigger.py
```

Figure 2-13 shows the wearable on the right wrist across the chest, and another wearable on the left wrist, with the palm upwards.

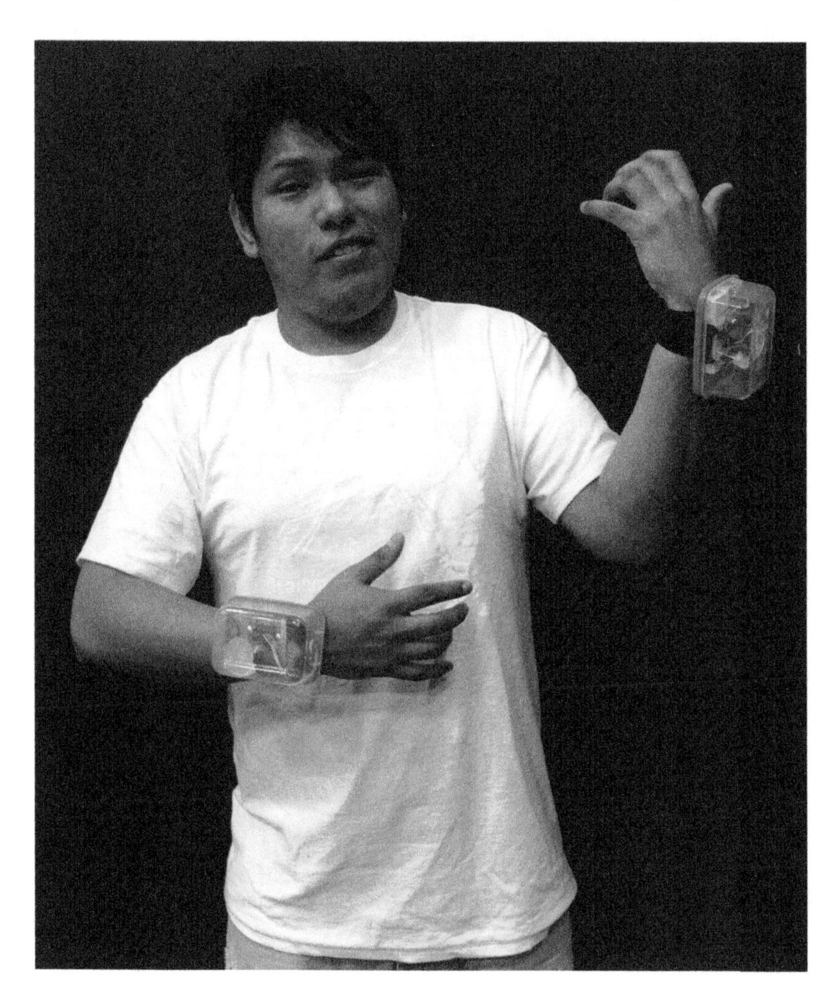

Figure 2-13. *Bass guitar*

Figure 2-14 shows the wearable on the right wrist across the chest, and another wearable on the left wrist with the palm backwards.

Figure 2-14. *Rhythm guitar*

Figure 2-15 shows the wearables on the right wrist and the left wrist, both palm downwards.

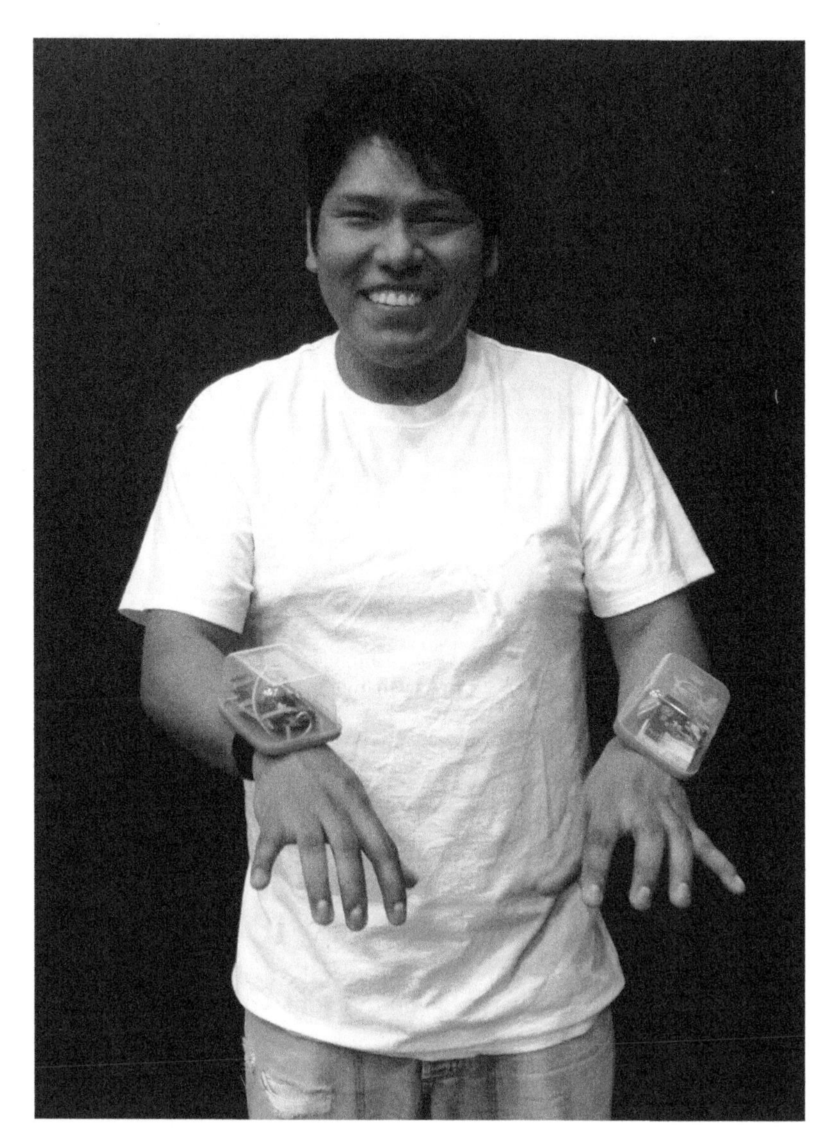

Figure 2-15. *Organ, electric piano*

Figure 2-16 shows the wearables on the left wrist and right wrist, each with palms angled 45 degrees down.

Figure 2-16. *Strings, French horns*

Figure 2-17 shows the wearable on the left wrist across the chest, palm backwards, and the wearable on the right wrist downward, palm backwards.

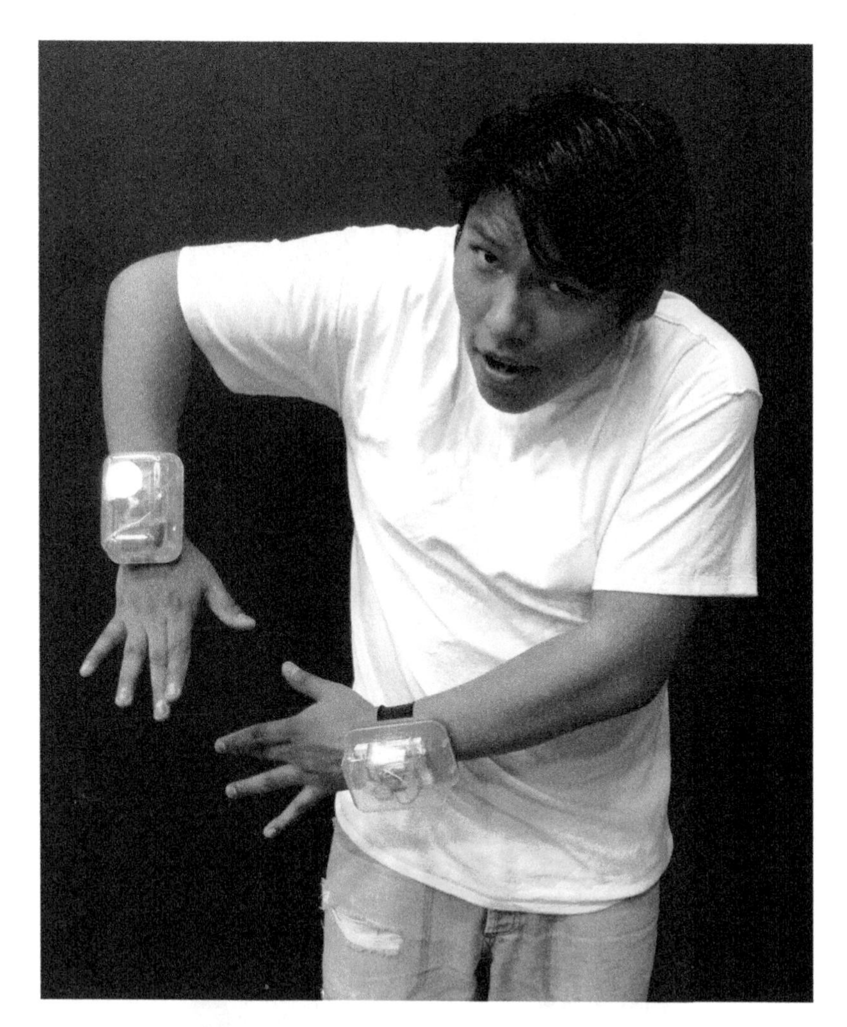

Figure 2-17. *Tambourine, congas*

Figure 2-18 shows the wearable on the right thigh at 45 degrees.

Figure 2-18. *Kit drums*

You can now produce the gesture-controlled One Man Band behavior. Instances of the `Synth` class that are created in *gesture_trigger.py* include paths that point to audio files.

Your First Performance

It's worth triggering each instrument and hearing the full multitrack recording several times before attempting a live performance. This will give you an idea of where the really exciting sections of each instrumental part can be found in the sequence and what combinations are likely to work. For conveniently previewing different instrument mixes of the Marvin Gaye recording, we added all of the tracks alongside each other in Jokosher, but there are many alternative multitrack editing environments for Linux, Mac, and Windows.

Your own One Man Band samples can be either *.wav* or *.oga* files (the formats supported by `pygame.mixer.Sound`). The current code assumes they are in a folder called *tracks*, inside of the *07_gesture_audio/pi* folder. Make sure the filenames in each synth entry in your program corresponds to one of the filenames in your samples. For playing the audio of the One Man Band, use headphones or an amplified speaker with a headphone jack on it.

Where Next?

We've used the Raspberry Pi to make Picussion, a platform for wireless, wearable audio applications. We've described how it can be a foundation for a range of interactions, games, and performances. The same build can be used in different ways, and additional inputs/outputs incorporated (such as vibration motors). For example, the online code repository includes a Fly Swat game that uses the same wearable design to provide a multiplayer game to "swat" and "splat" whichever wearable is flashing its LED. So where to go from here?

When you feel comfortable with all of the technical aspects of Picussion, we suggest approaching new applications from the perspective of outcome. Who is your audience? Where will your design be used? What messages do you want to communicate? What affordances are appropriate? What values are important within your design?

We sincerely hope you have enjoyed building this project, and welcome any feedback. Let the experimentation begin!

Raspberries from Scratch

3

By Sjoerd Dirk Meijer

You can create your own conductive keyboard out of everyday objects and a Raspberry Pi (Figure 3-1). Follow along in this chapter to learn how to program two games in Scratch, make a homemade conductive keyboard with edible raspberries, and then use your keyboard to play the games.

Figure 3-1. *A conductive keyboard*

You don't need to be an electronics guru to complete this project. It's helpful if you know the difference between a resistor and a capacitator, but it's not essential. This chapter will provide you with everything you need to complete this project, including:

- Install and configure all of the necessary software on the Raspberry Pi.
- Make a conductive keyboard out of a solderless breadboard (or optionally on a stripboard).
- Code two Scratch projects: an art project and a memory game.

Let's start building!

The Conductive Keyboard

Here's a list of the components you need to build the conductive keyboard.

Materials

A list of necessary materials is provided in Table 3-1.

Table 3-1. *Bill of materials*

Item	Item
1x ATmega328P-PU (DIP28) or ATmega8 (DIP28)	1x 10 uF capacitor
3x 100 nF/0.1 uF capacitor	2x 22 pF capacitor
16x 22 MΩ or 20 MΩ resistor (1/4W)	1x 10 kΩ resistor (1/4W)
1x 2.2 kΩ resistor (1/4W)	2x 68 Ω resistor (1/4W)
5x 330 Ω resistor (1/4W)	5x LED (5mm) (4 different colors)
1x 1N4148 diode	2x 3.6V Zener diode (max. 0.5W!)
1x 16 MHz crystal	2x button
3x mini breadboard (170 holes)	1x 40-pin single row male pin header strip (with 2.54 mm/0.1 inch spacing)
Jumper wires male-male	Jumper wires male-female

Item	Item
Raspberry Pi (this project has been tested on models B and B+)	1x USB cable (with male A connector)

Resistors

The value of a resistor is coded in four colored bars. The resistors you need for this project have the following colors:

- 22 MΩ: red-red-blue-gold or 20 MΩ: red-black-blue-gold
- 10 kΩ: brown-black-orange-gold
- 2.2 kΩ: red-red-red-gold
- 330 Ω: orange-orange-brown-gold
- 68 Ω: blue-gray-black-gold

Although we walk you through the solderless breadboard version of this project, you may want to use a stripboard, which requires soldering and will make your keyboard more durable. Table 3-2 lists the optional materials needed if you go this route.

Table 3-2. *Optional materials*

1x stripboard (94 x 55 mm)
1x 28 pin DIP IC socket
Alligator clips
ESD Anti Static Wrist Strap

Installing Software

To begin, you need to configure the Raspberry Pi and install some software. This can be done in two ways:

Automatically on a clean install of Raspbian
: This is the easiest way of installing all of the necessary software. With four commands, everything is installed and configured.

Manually on a clean or existing install of Raspbian
: If you're familiar with Raspbian, you can go this route. All of the required commands are covered in Appendix A, and if necessary you can adapt these commands to your installation.

We'll continue with the automatic install.

Automatic installation

Prepare an SD card with Raspbian. Instructions for this can be found on the Raspberry Pi website (*http://bit.ly/1oalhoW*).

1. Insert the SD card into the Raspberry Pi and connect your Pi to a power source.
2. When your Pi starts up, a program called "raspi-config" will run. Use the Tab and arrow keys to go to the Finish button, and press Enter.
3. Type the following case-sensitive commands at the prompt, pressing the Enter key after each line:

```
wget http://fse.link/r08
unzip r08
chmod +x fSE.sh
./fSE.sh
```

These commands will download, compile, and install all of the necessary software required for this project.

Raspberry Pi as AVR Programmer

Our conductive keyboard uses an ATmega chip, which needs a bootloader to run. A new ATmega chip doesn't come preinstalled with a bootloader, so we must manually add it. There are special devices, called AVR Programmers, that are used for adding the bootloader, but for this project we will use the Raspberry Pi to burn the bootloader to our ATmega chip.

You now have all of the necessary software installed, including the software needed to burn the bootloader, called AVRDUDE. With AVRDUDE it's possible to burn *.hex* files containing the bootloader to the chip. The necessary *.hex* files, based on USBaspLoader (*https://*

github.com/baerwolf/USBaspLoader), are provided in the conductive keyboard's software package.

To burn the bootloader, the following materials are needed:

- 1x ATmega328P-PU (DIP28) or ATmega8 (DIP28)
- 2x mini breadboard (170 holes)
- 4x 330 Ω resistor (1/4W)
- 1x 16000 MHz crystal
- 2x 22 pF capacitor (1/4W)
- 8x jumper wire

Be careful with the legs of the ATmega chip. If they splay out too much, ease them into position by gently pressing one side of the chip against a tabletop (14 pins at a time). The chip should slide into the holes of the breadboard easily.

Now let's place the components:

1. First, lock the two breadboards together.

Make sure the ATmega chip is placed correctly. There's a notch at one end of the chip. Place the notch in the same position as shown in the schematics in Figure 3-2, otherwise you could damage your chip.

2. Connect all of the components in this order:

Component	From	To
ATmega328P-PU	Left J:4	Right A:4
330 Ω resistor	Left E:4	Left F:4
330 Ω resistor	Right E:13	Right F:13

Component	From	To
330 Ω resistor	Right E:14	Right F:14
330 Ω resistor	Right E:15	Right F:15
16000 MHz crystal	Left F:12	Left G:13
22 pF capacitor	Left H:11	Left H:12
22 pF capacitor	Left I:11	Left I:13
wire	Left I:10	Right B:12
wire	Left I:11	Right B:10
wire	Right J:13	RPi SCLK
wire	Right J:14	RPi MISO
wire	Right J:15	RPi MOSI
wire	Left A:4	RPi GPIO25
wire	Left F:11	RPi GND
wire	Left F:10	RPi 3V3

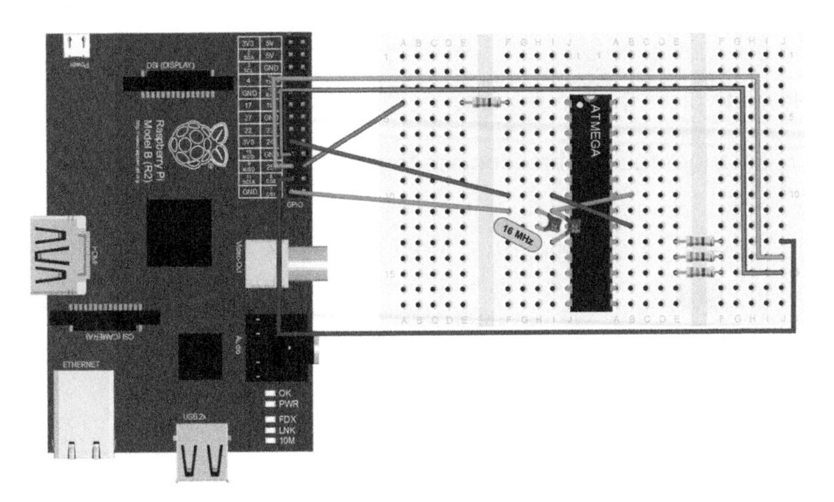

Figure 3-2. *AVR Programmer schematics*

Wiring on the Raspberry Pi B+

The wiring for a Raspberry Pi B+ is the same as for the Raspberry Pi B, even though the B+ has 12 more pins. Just connect the wires as shown in Figure 3-2. The 12 extra pins on the B+ are below the green wire and aren't used in this section.

Your Raspberry Pi should be running and connected to the Internet (see Appendix A for connecting to WiFi) and to an HDMI monitor or TV. You can now begin the process of burning the bootloader onto the ATmega chip. Launch the LXTerminal from the Start menu if you're in the graphical user interface of Raspbian, or if you're already on the command line, then you are ready to begin.

When you're at the command-line prompt, you'll have to type in two commands to start burning the bootloader.

For ATmega328P-PU chips, type and execute the following commands:

```
> cd /home/pi
> ./atm328p.sh
```

For ATmega8 chips, type and execute the following commands:

```
> cd /home/pi
> ./atm8.sh
```

If everything goes well, you'll get an output that looks like the following:

```
avrdude: AVR device initialized and ready to accept instructions

Reading | ################################################## | 100% 0.00s

avrdude: Device signature = 0x1e950f
avrdude: NOTE: "flash" memory has been specified, an erase cycle will be
performed
         To disable this feature, specify the -D option.
avrdude: erasing chip
avrdude: reading input file "ShrimpUSB_rev1_atmega328p_16MHz.hex"
avrdude: input file ShrimpUSB_rev1_atmega328p_16MHz.hex auto detected as
Intel Hex
avrdude: writing flash (30880 bytes):
```

```
Writing | ################################################## | 100% 0.89s

avrdude: 30880 bytes of flash written
avrdude: verifying flash memory against
ShrimpUSB_rev1_atmega328p_16MHz.hex:
avrdude: load data flash data from input file
ShrimpUSB_rev1_atmega328p_16MHz.hex:
avrdude: input file ShrimpUSB_rev1_atmega328p_16MHz.hex auto detected as
Intel Hex
avrdude: input file ShrimpUSB_rev1_atmega328p_16MHz.hex contains 30880
bytes
avrdude: reading on-chip flash data:

Reading | ################################################## | 100% 0.73s

avrdude: verifying ...
avrdude: 30880 bytes of flash verified

avrdude: safemode: Fuses OK (E:04, H:D0, L:D7)

avrdude done.  Thank you.


avrdude: AVR device initialized and ready to accept instructions

Reading | ################################################## | 100% 0.00s

avrdude: Device signature = 0x1e950f
avrdude: reading input file "0xD7"
avrdude: writing lfuse (1 bytes):

Writing | ################################################## | 100% 0.00s

avrdude: 1 bytes of lfuse written
avrdude: verifying lfuse memory against 0xD7:
avrdude: load data lfuse data from input file 0xD7:
avrdude: input file 0xD7 contains 1 bytes
avrdude: reading on-chip lfuse data:

Reading | ################################################## | 100% 0.00s

avrdude: verifying ...
avrdude: 1 bytes of lfuse verified
avrdude: reading input file "0xD0"
avrdude: writing hfuse (1 bytes):

Writing | ################################################## | 100% 0.00s

avrdude: 1 bytes of hfuse written
avrdude: verifying hfuse memory against 0xD0:
avrdude: load data hfuse data from input file 0xD0:
avrdude: input file 0xD0 contains 1 bytes
avrdude: reading on-chip hfuse data:

Reading | ################################################## | 100% 0.00s
```

```
avrdude: verifying ...
avrdude: 1 bytes of hfuse verified
avrdude: reading input file "0x04"
avrdude: writing efuse (1 bytes):

Writing | ################################################## | 100% 0.00s

avrdude: 1 bytes of efuse written
avrdude: verifying efuse memory against 0x04:
avrdude: load data efuse data from input file 0x04:
avrdude: input file 0x04 contains 1 bytes
avrdude: reading on-chip efuse data:

Reading | ################################################## | 100% 0.00s

avrdude: verifying ...
avrdude: 1 bytes of efuse verified

avrdude: safemode: Fuses OK (E:04, H:D0, L:D7)

avrdude done.  Thank you.
```

If you receive output like the following, then please recheck your wiring:

```
avrdude: error: AVR device not responding
avrdude: initialization failed, rc=-1
         Double check connections and try again, or use -F to override
         this check.


avrdude done.  Thank you.
```

You've now burned the bootloader onto the ATmega chip. You need to remove some of the components on the breadboard that were used to burn the bootloader before continuing. Remove all of the wires, except for the two wires that cross the ATmega chip. Then, remove the resistors. The rest of the components remain in place (see Figure 3-3 for an illustration of what your board should look like).

Building the Conductive Keyboard

Now that the ATmega chip contains the bootloader, it is time to add all of the remaining components for the conductive keyboard. These components will be used with our Scratch games.

Exercise 1

Be sure you have removed the excess components from the bootloading process and that your board looks like the one shown in Figure 3-3. As shown in the figure, the ATmega chip is the heart of the conductive keyboard, controlling its required inputs, outputs, and triggers.

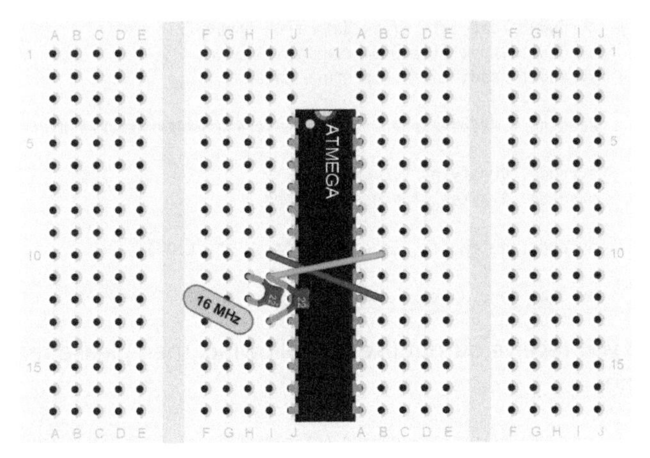

Figure 3-3. *Exercise 1*

The crystal is the clock, keeping the conductive keyboard running. The capacitors are for protection: they store energy and release it steadily so there aren't any energy spikes.

Exercise 2

Add the following components to the breadboard as shown in Figure 3-4 and the following chart. You can add the components in any order, but it's easiest to add the wires last.

- 3x 100 nF capacitor (1/4W)
- 1x 10 uF capacitor (1/4W)
- 1x 1N4148 diode
- 1x 10 kΩ resistor (1/4W)
- 1x button
- 3x jumper wire

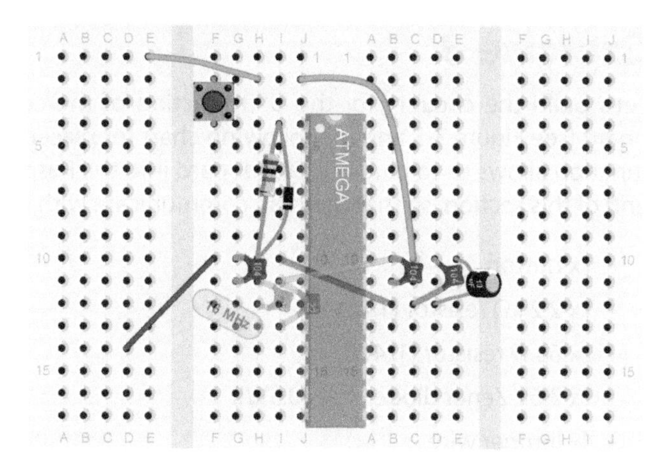

Figure 3-4. *Exercise 2*

Component	From	To
100 nF capacitor	Left G:10	Left G:11
100 nF capacitor	Right B:10	Right B:11
100 nF capacitor	Right D:10	Right C:12
10 uF capacitor	Right E:10	Right D:12
1N4148 diode	Left I:4	Left H:10
10 kΩ resistor	Left I:4	Left H:10
button	Left F:2	Left F:4
wire	Left E:1	Left H:2
wire	Left J:2	Right C:10
wire	Left F:10	Left D:14

The diode, resistor, and the button build the reset circuitry. The capacitors have the same function as those already in place.

The large capacitor (10 uF) has polarity. Make sure that the white stripe is pointing in the same direction as in the image. The diode has a line on it, so make sure this line is on the same side as in the image.

Exercise 3

Let's build the circuitry for the USB function of the conductive keyboard (see Figure 3-5 and the following chart for placement). The USB function allows us to plug our breadboard into the Raspberry Pi at the end of this section, so that they can communicate with each other.

- 1x button
- 1x 2.2 kΩ resistor (1/4W)
- 2x 68 Ω resistor (1/4W)
- 2x 3.6V Zener diode (max. 0.5W!)
- 4x jumper wire

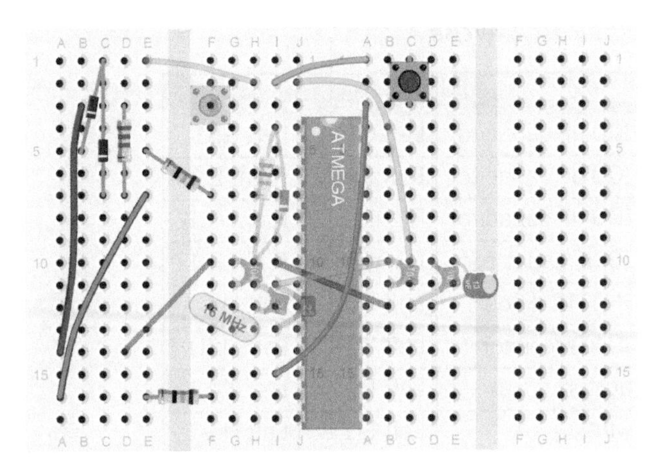

Figure 3-5. *Exercise 3*

Component	From	To
button	Right C:1	Right C:3
2.2 kΩ resistor	Left D:3	Left D:7
68 Ω resistor	Left E:5	Left F:7
68 Ω resistor	Left E:16	Left F:16
3.6V Zener diode	Left C:1	Left B:5
3.6V Zener diode	Left C:1	Left C:7
wire	Left I:2	Right A:1
wire	Left I:15	Right A:3

Component	From	To
wire	Left B:3	Left A:14
wire	Left E:7	Left A:16

When the button placed on C:1/C:3 is pressed during the booting or resetting of the conductive keyboard, it will place the ATmega in programming mode in order to upload software.

Exercise 4

Connect the conductive keyboard to the Raspberry Pi. You use a USB cable for this.

Since you are using a breadboard, you can't use a female USB connector (it won't fit). So you need to cut the wires of the USB cable below the connector, leaving the USB-A connector on the cable! Then expose the wires inside it so that you can connect them directly into the breadboard. Most USB cables contain a white, green, black, and red wire:

- The white cable is DATA–.
- The green cable is DATA+.
- The black cable is GROUND.
- The red wire is +5V.

Go ahead and connect the USB cable to the breadboard as shown in Figure 3-6 and the following chart:

- 1x USB cable
- 1x strip of 7 male pin headers

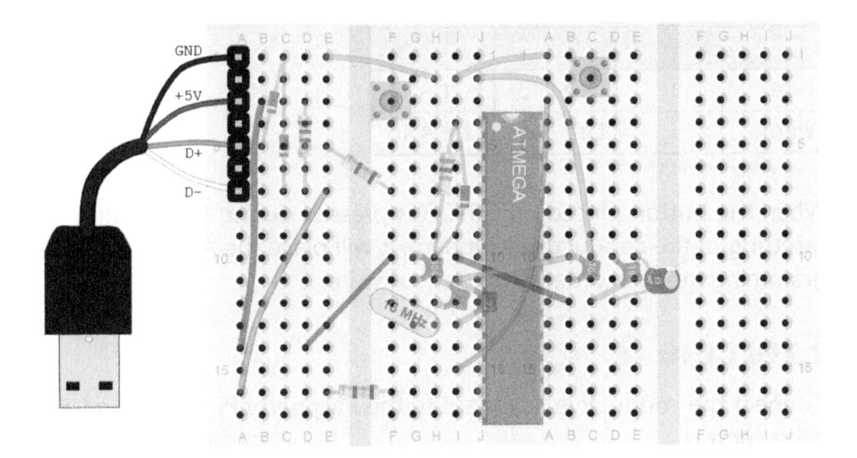

Figure 3-6. *Exercise 4*

Component	From	To
USB cable GND (black)	cable	Left A:1
USB cable GND (red)	cable	Left A:3
USB cable GND (green)	cable	Left A:5
USB cable GND (white)	cable	Left A:7
pin headers	Left A:1	Left A:7

Strip the four wires and place them into the corresponding holes of the breadboard. To strengthen the connection, also place a strip of male pin headers into the same holes. Be sure to leave the four wires long enough to allow the pin headers to push the ends of the wires into the breadboard (Figure 3-7).

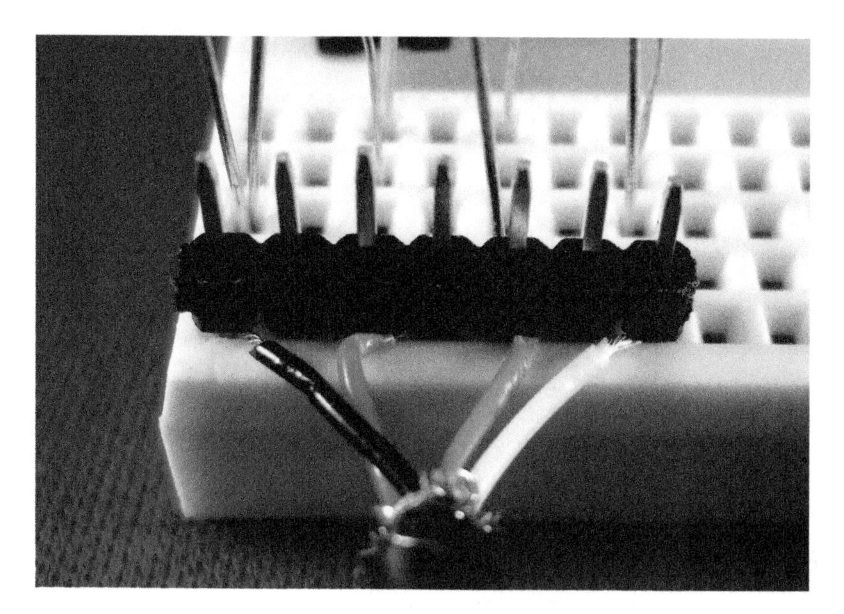

Figure 3-7. *USB cable*

Exercise 5

Add an LED to the breadboard (see Figure 3-8 and the following chart). The LED is perfect for testing the conductive keyboard, and it is used later as an indicator light.

- 1x LED (5mm)
- 1x 330 Ω resistor (1/4W)
- 1x jumper wire

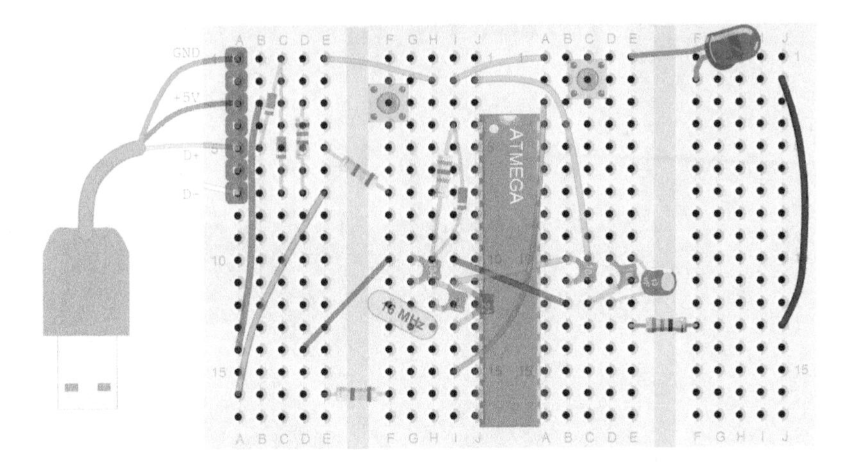

Figure 3-8. *Exercise 5*

Component	From	To
LED	Right E:1	Right F:2
330 Ω resistor	Right E:13	Right F:13
wire	Right J:2	Right J:13

Make sure that the flat side of the LED (with the short leg) is in the right breadboard at E:1.

Testing the Conductive Keyboard

Let's take a moment to test the keyboard and make sure all of the current wiring is correct.

Your breadboard should now look like the one in Figure 3-9.

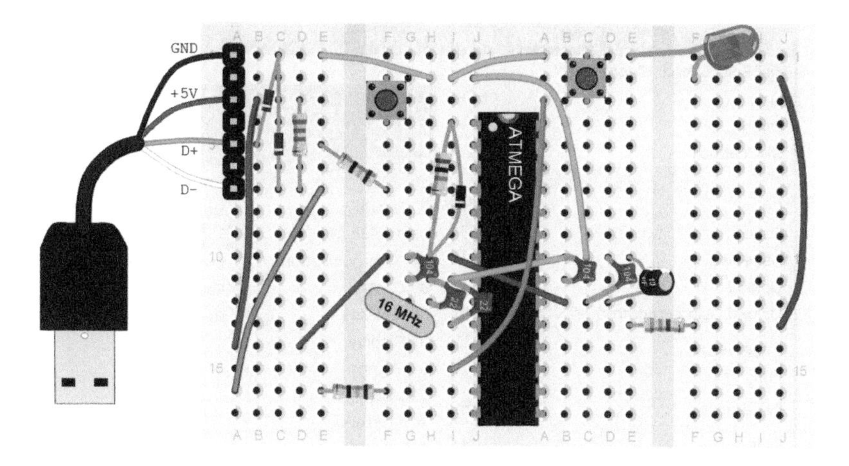

Figure 3-9. *The finished conductive keyboard USB*

Test the conductive keyboard by starting up Raspbian on your Raspberry Pi, and then launch the Arduino IDE from the Start menu, as shown in Figure 3-10.

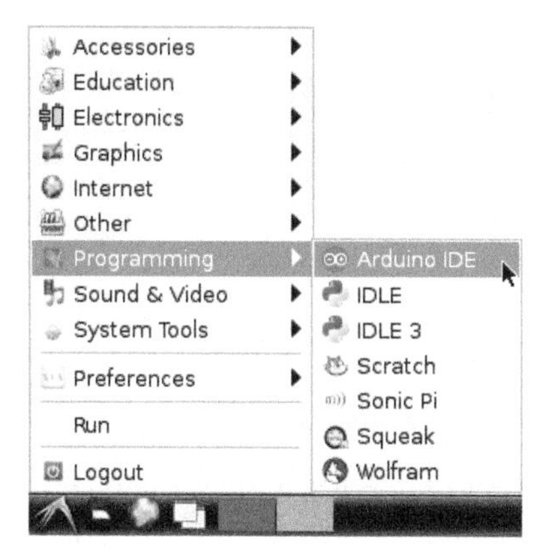

Figure 3-10. *Arduino IDE*

The Arduino IDE contains a lot of example sketches. A sketch is a software program that you will soon be loading on to the ATmega chip.

Use the Blink sketch to run the test. The Blink sketch should blink the LED on the breadboard continuously (1 second on, 1 second off), displaying a working project.

Go to File → Examples → 01. Basics → Blink, as shown in Figure 3-11.

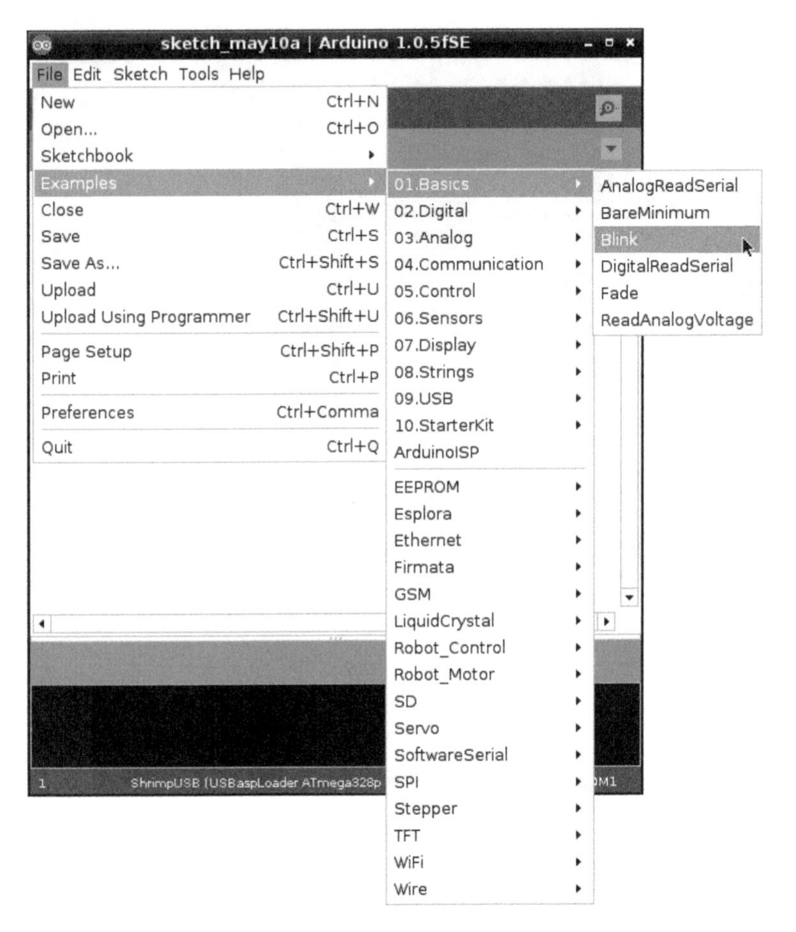

Figure 3-11. *Opening the Blink sketch*

At the bottom of the Arduino IDE window, you should see ShrimpUSB (the breadboard's name), along with the ATmega chip name. If a different board or chip is shown in your Arduino IDE, please go to Extra → Board and choose ShrimpKey. Also check that USBasp is selected under Extra → Programmer (this software simulates a USB chip, which is explained in Appendix A). Verifying these settings will ensure the successful execution of the Blink sketch.

Be sure to also check the Arduino IDE's version number. It should end with "fSE," otherwise the sketch you'll be using later won't work (see Appendix A for more details).

Now connect the conductive keyboard to a USB port on the Raspberry Pi with the USB cable. Hold the programming button (C:1/C:3) while connecting the USB cable. You can release the button as soon as you've connected the cable.

Open an LXTerminal session and type:

```
> lsusb
```

The output should contain something like this (the numbers can be different):

```
Bus 001 Device 064: ID 16c0:05dc VOTI shared ID for use with libusb
```

If the output is not correct, hold the programming button (C:1/C:3) on the conductive keyboard, press the reset button (F:2/F:4), and release the programming button after 2–3 seconds. Check the output again after typing:

```
> lsusb
```

If the proper output is still not showing up, a component on the breadboard may be in the wrong place. Go back and check your wiring.

Assuming you see the desired output, you can upload the Blink sketch on to the ATmega chip. In the Arduino IDE, click the round button with an arrow. The sketch is compiled and then uploaded to the ATmega chip.

After uploading the sketch, press the reset button on the conductive keyboard to run the Blink software. The LED should now blink with an interval of 1 second and we are ready to extend our keyboard.

Extending the Conductive Keyboard

Now it is time to finish the conductive keyboard so that it can interact with the Scratch games.

The conductive keyboard uses 16 of its pins as input. All of these pins are connected to a pull-up resistor with a value of 22 MΩ. Pull-up means that the resistors are connected between +5V and the input pins, so there's a constant current on the pins. Connecting a conductive object (like a raspberry) to that pin enables the current to flow into that object. When you connect yourself to the GROUND pin (like the girl does in Figure 3-12) and touch the object, the current is drawn away. The conductive keyboard detects this change in current. At that time, a signal is sent to the connected computer, transmitting the sent key.

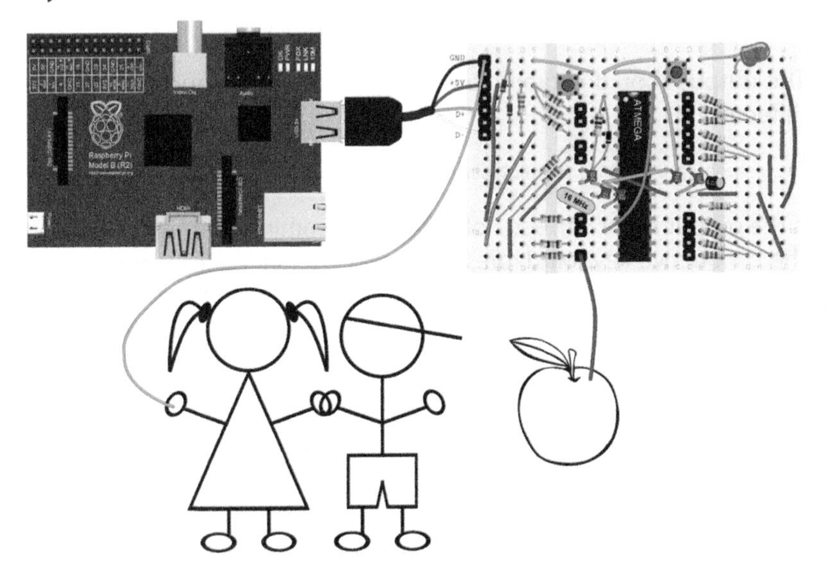

Figure 3-12. *Usage overview*

Exercise 6

Let's start at the right side of the conductive keyboard. Connect 10 of the 22 MΩ resistors (1/4W) into the conductive keyboard as shown in Figure 3-13 and the chart that follows.

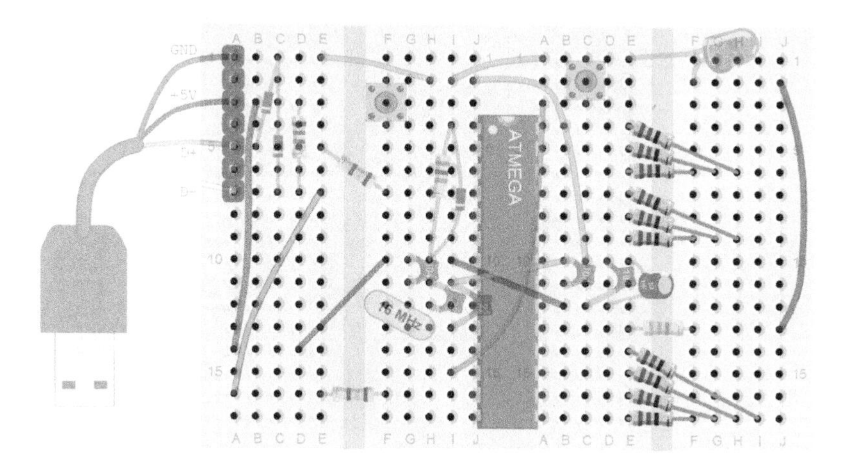

Figure 3-13. *Exercise 6*

Component	From	To
22 MΩ resistor	Right E:4	Right H:6
22 MΩ resistor	Right E:5	Right G:6
22 MΩ resistor	Right E:6	Right F:6
22 MΩ resistor	Right E:7	Right H:9
22 MΩ resistor	Right E:8	Right G:9
22 MΩ resistor	Right E:9	Right F:9
22 MΩ resistor	Right E:14	Right I:17
22 MΩ resistor	Right E:15	Right H:17
22 MΩ resistor	Right E:16	Right G:17
22 MΩ resistor	Right E:17	Right F:17

Exercise 7

Now, let's power up these resistors with four jumper wires. Connect the resistors to +5V as as shown in Figure 3-14 and the chart that follows.

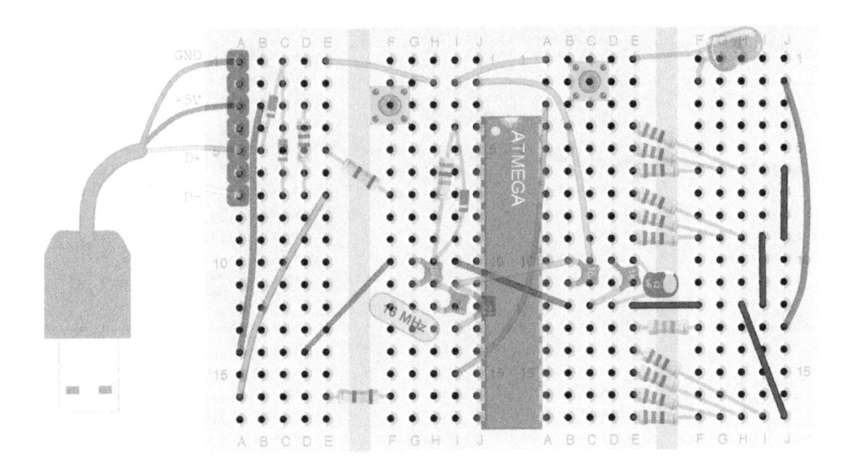

Figure 3-14. *Exercise 7*

Component	From	To
wire	Right E:12	Right F:12
wire	Right J:6	Right J:9
wire	Right I:9	Right I:12
wire	Right H:12	Right J:17

Exercise 8

Place six of the 22 MΩ resistors (1/4W) and one jumper wire into the breadboard holes on the left side of the conductive keyboard (see Figure 3-15 and the following chart).

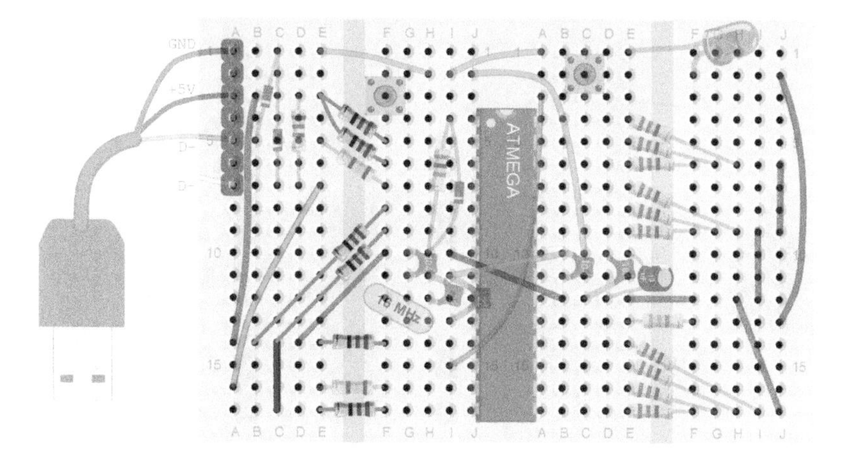

Figure 3-15. *Exercise 8*

Component	From	To
22 MΩ resistor	Left E:3	Left F:5
22 MΩ resistor	Left E:3	Left F:6
22 MΩ resistor	Left B:14	Left F:8
22 MΩ resistor	Left C:14	Left F:9
22 MΩ resistor	Left E:14	Left F:14
22 MΩ resistor	Left E:17	Left F:17
wire	Left C:14	Left C:17

Exercise 9

Add the following pin headers to the breadboard (see Figure 3-16 and the following chart):

- 1x strip of 1 male pin headers
- 3x strip of 2 male pin headers
- 1x strip of 4 male pin headers
- 1x strip of 6 male pin headers

The headers will be used to connect your objects to the conductive keyboard keys.

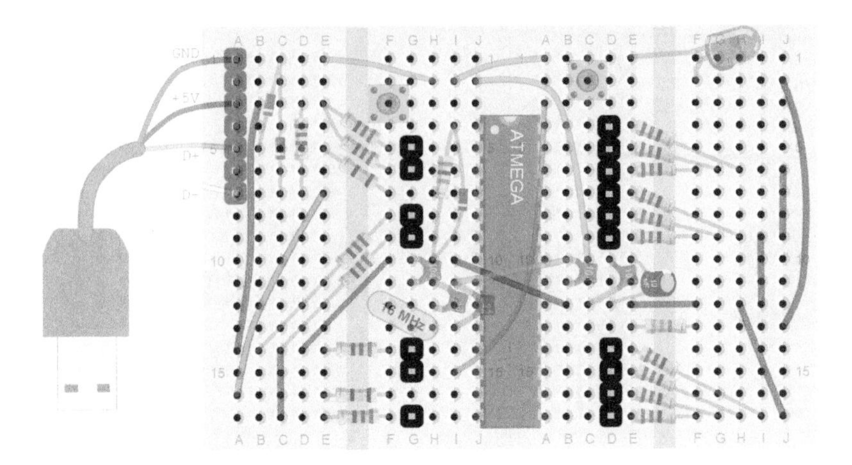

Figure 3-16. *Exercise 9*

Component	From	To
pin header (1)	Left G:17	
pin header (2)	Left G:5	Left G:6
pin header (2)	Left G:8	Left G:9
pin header (2)	Left G:14	Left G:15
pin header (4)	Right D:14	Right D:17
pin header (6)	Right D:4	Right D:9

The hardware portion of the conductive keyboard is now complete and should look like Figure 3-17.

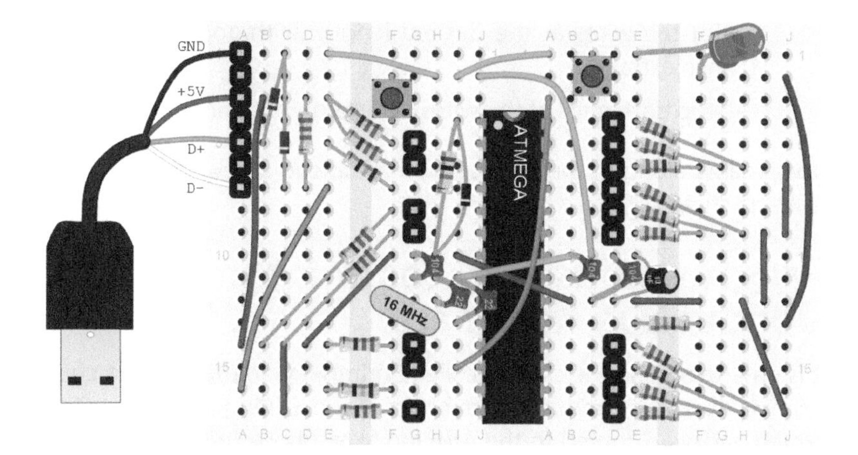

Figure 3-17. *Finished conductive keyboard*

And Figure 3-18 is an actual photo of the finished conductive keyboard.

Figure 3-18. *Photo of a finished conductive keyboard*

If you want, you can now make the breadboard tidier:

- Cut the legs of the components, so that they're not too long.
- Use normal wires (with a solid core).

— Cut the wires at the desired length (with approx. 1 cm/0.4 inch extra).

— Strip both ends of the wires (at approx. 0.5 cm/0.2 inch).

Now, follow along in the next section to upload a new sketch to the ATmega chip.

Programming and Testing the Conductive Keyboard

It is time to upload the sketch (called ShrimpKey) to the ATmega chip. This will allow the conductive keyboard to run. During installation, the sketch has been copied to the Arduino Sketchbook (running in Raspbian on the Raspberry Pi), making it easily available from within the Arduino IDE.

Go to File → Sketchbook → ShrimpKey (Figure 3-19).

The ShrimpKey sketch is now loaded. Now, let's upload it to the ATmega chip:

1. Connect the conductive keyboard with the USB cable to the running Raspberry Pi while holding the programming button (C:1/C:3).
2. Release the programming button as soon as you've connected the USB cable, just as you did with the Blink sketch earlier in this chapter.
3. In the Arduino IDE, click the upload button.

The sketch is uploaded when you see Figure 3-20. Ignore the error message about the program size, since it doesn't impact the quality of the uploaded sketch.

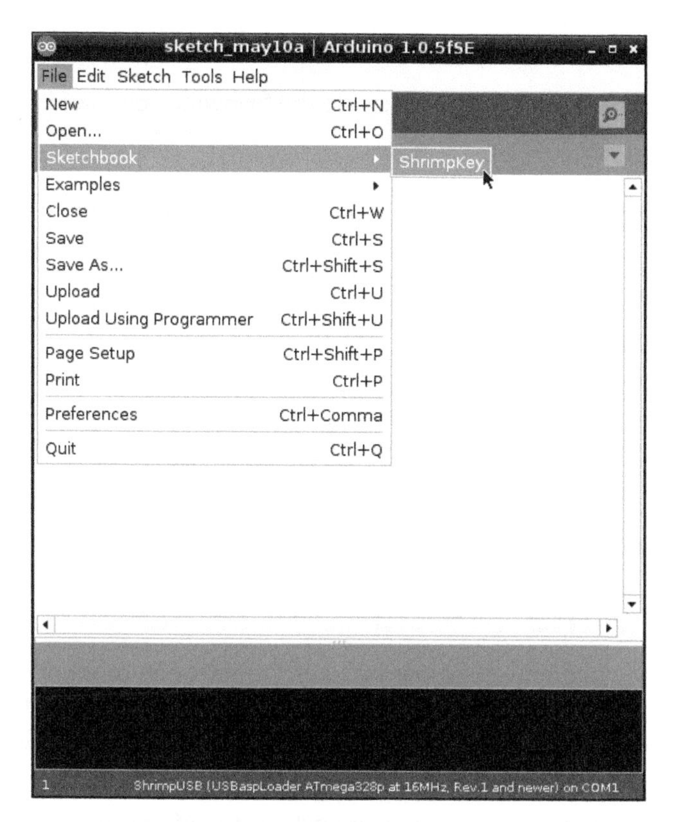

Figure 3-19. *Opening the ShrimpKey sketch*

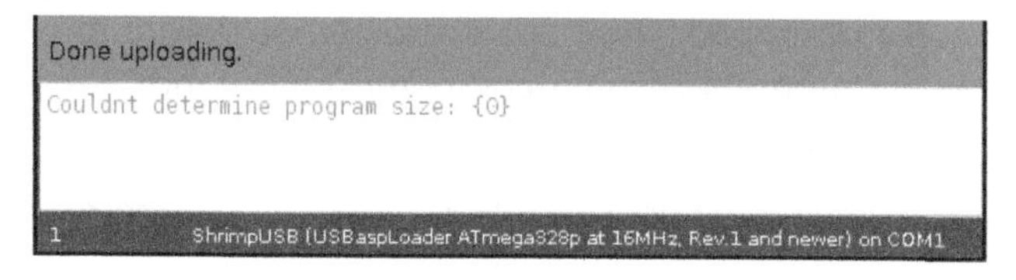

Figure 3-20. *Output*

Press the reset button (F:2/F:4) on the conductive keyboard to run the software. The LED should blink a couple of times.

Open the LXTerminal session and execute the following command:

```
> lsusb
```

The output should contain something like this (the numbers can be different):

```
Bus 001 Device 082: ID 4242:e131 USB Design by Example
```

To be sure everything works, execute the following command:

```
> dmesg
```

At the end of the output, you should see something like this:

```
[29015.813344] usb 1-1.3: New USB device found, idVendor=4242,
idProduct=e131
[29015.813385] usb 1-1.3: New USB device strings: Mfr=1, Product=2,
SerialNumber=0
[29015.813406] usb 1-1.3: Product: ShrimpKey
[29015.813424] usb 1-1.3: Manufacturer: fromScratchEd.nl
[29015.835997] input: fromScratchEd.nl ShrimpKey as /devices/platform/
bcm2708_usb/usb1/1-1/1-1.3/1-1.3:1.0/input/input32
[29015.838112] hid-generic 0003:4242:E131.0021: input,hidraw0: USB HID
v1.01 Mouse [fromScratchEd.nl ShrimpKey] on usb-bcm2708_usb-1.3/input0
```

Congratulations, you've made your own conductive keyboard!

Using the Conductive Keyboard

Using the conductive keyboard is easy. Connect one or more objects (they have to be a little conductive, so rubber or plastics won't work) to the pin headers. Connect the conductive keyboard to the USB port. You can now start typing.

Typing? Well yes, the conductive keyboard is of course a keyboard (and a mouse). You can use it as a keyboard for your favorite game or you can make drum pads from flowers to compose your own music. As long as the program on your computer understands key presses, you can use the conductive keyboard to interact with the program.

To see how it works, start up a web browser on the Raspberry Pi:

1. Go to a long web page (i.e., *raspberrypi.org*).
2. Scroll down this web page completely.
3. Connect a conductive object to the bottom-left pin header on the conductive keyboard.
4. Connect yourself to the GROUND pin of the conductive keyboard.
5. Connect the USB cable to the Raspberry Pi (see Figure 3-21).

Figure 3-21. *Final wiring*

6. Now tap the conductive object and the web page should scroll up!

There's a small (unharmful) current passing through your body when you touch the object. The conductive keyboard detects that and sends the key press to the Raspberry Pi.

Alligator Clips

To easily connect objects to your conductive keyboard, connect a jumper wire to the pin header. Use alligator clips to connect the object and the jumper wire. Alligator clips will make a better connection to most objects and can also easily be clipped on to the jumper wire. This method also extends the length of your wire.

Table 3-3 presents your key assignments, so that you can start using your conductive keyboard. The pin headers are mapped with the keys as shown in the table and Figure A-4 ([k]eyboard, [m]ouse).

Table 3-3. *Key assignment*

Left side	Pin	Right side	Pin
		Shift [k]	19
a [k]	0	Move mouse pointer right [m]	18
s [k]	1	Move mouse pointer left [m]	17
		Move mouse pointer down [m]	16
d [k]	3	Move mouse pointer up [m]	15
w [k]	4	Right mouse click [m]	14
Arrow left [k]	5	Arrow right [k]	12
(output)	6	Arrow down [k]	11
		Space [k]	10
Arrow up [k]	8	Left mouse click [m]	9

Feel free to change the order of the keys (this is explained in Appendix A).

ESD Anti Static Wrist Strap

For children it's easier to use both hands to play with the objects connected to the conductive keyboard. They don't want one hand holding the ground wire so that they have only one hand free to play. Connect a jumper wire to the GND wire of the USB cable and connect an ESD Anti Static Wrist Strap to the jumper wire. Put the wrist strap on to the arm of the child and secure it. This will give children the freedom of movement they need to play with the conductive keyboard.

It's now time to create your Scratch games so that you can put your conductive keyboard to use.

Using the ShrimpKey with Scratch

In this section, we're going to use the ShrimpKey with Scratch. "With Scratch, you can program your own interactive stories, games, and animations—and share your creations with others in the online community. Scratch helps young people learn to think creatively, reason systematically, and work collaboratively—essential skills for life in the twenty-first century. Scratch is a project of the Lifelong Kindergarten Group at the MIT Media Lab. It is provided free of charge," as written on the Scratch website (*http://fse.link/r17*).

The ShrimpKey is a perfect companion for Scratch, because you can make your own program (i.e., game, interactive story, art) in Scratch and control it with a controller you've created with the ShrimpKey. You can use PlayDoh, fruit, vegetables, sauces, flowers, people, or any other conductive material to make this controller. Using the ShrimpKey in Scratch is very easy because the ShrimpKey outputs key presses and Scratch can detect these.

In this section, we're going to make two projects in Scratch. The first one is an interactive art project ("Project 1: Interactive Art Project" on page 102). The second project is a game called "Simon Game," which also uses the GPIO pins of the Raspberry Pi ("Project 2: Memory Game" on page 110).

But I'll start with explaining what Scratch is and how it works.

A Scratch Primer

Before we create our Scratch games, let's go over the basics. Scratch is a graphical programming language. You can see its user interface in Figure 3-22. Unlike other textual programming languages that have a strict syntax, the code in Scratch is written on blocks (Figure 3-23).

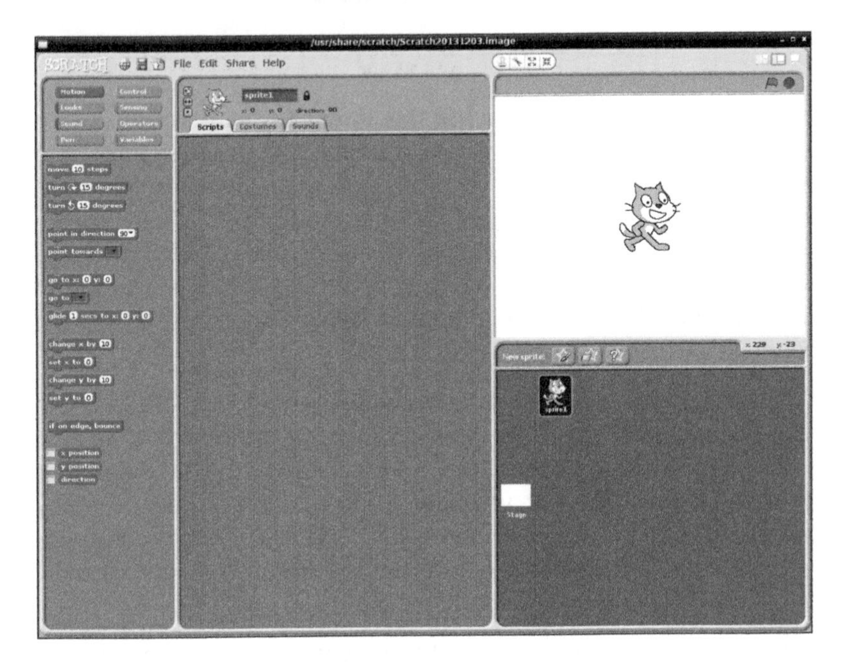

Figure 3-22. *The user interface of Scratch*

With Scratch, programmers can concentrate on learning to think like a programmer, without being distracted on the correct usage of a written programming language. By right-clicking a block, you'll get a "help" option. If you click it, you'll get a pop-up in which the usage of that block is explained.

Figure 3-23. *A block*

All of the blocks are grouped by categories and have their own color. The Scratch categories are shown in Figure 3-24.

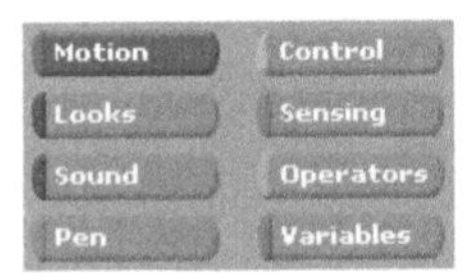

Figure 3-24. *The categories*

You can code just by *dragging* the block outside of the category palette and into the scripts zone (the middle section of the screen).

The code blocks can be snapped together like building blocks, and a script (Figure 3-25) can be executed by clicking the green flag.

Figure 3-25. *A script*

A Scratch project is based on sprites. Sprites are objects that are placed on the stage (the top-right part of the screen). The stage is divided into coordinates, based on an x-axis and a y-axis. Sprites can be placed on the stage with an *x*- and *y*-coordinate (Figure 3-26).

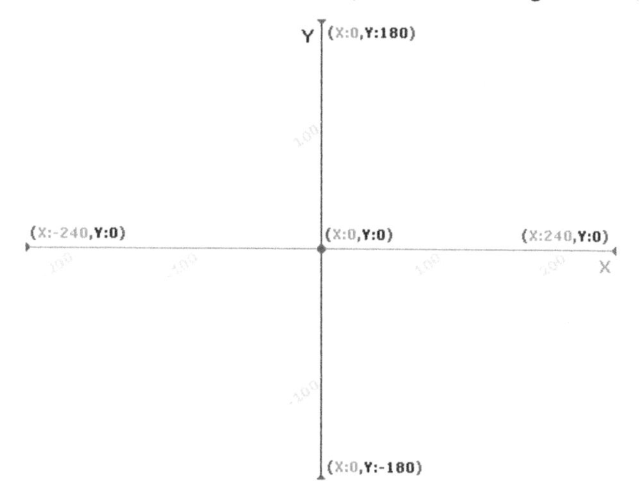

Figure 3-26. *Coordinate grid with x- and y-axes*

Every sprite has it's own code. It's possible, however, to broadcast messages between sprites and let them interact with each other.

Scratch 1.4 is installed by default on Raspbian and works well with the Raspberry Pi. Note that a new version of Scratch, Scratch 2.0, has more features than Scratch 1.4 and was released in May 2013. Unfortunately, Scratch 2.0 is made in Flash and therefore it can't run on the Pi. Given this fact, the first Scratch project where you will create artwork in this chapter uses Scratch 1.4 on the Raspberry Pi. The second project, the memory game, will use ScratchGPIO, which was installed earlier in this chapter.

Using the Conductive Keyboard in Scratch

The blocks in Figure 3-27 can be used to integrate the conductive keyboard with your Scratch project.

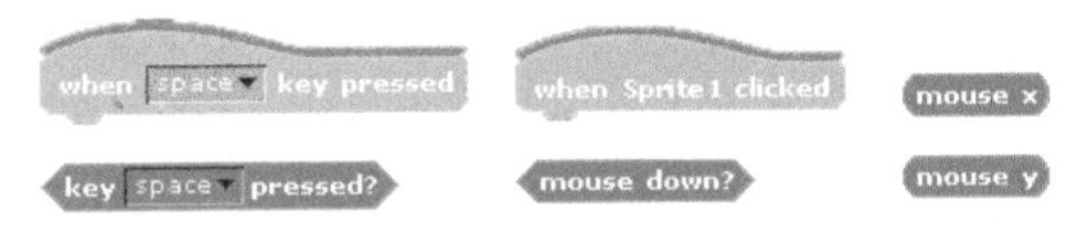

Figure 3-27. *Conductive keyboard blocks*

The *when space key pressed* block is called a hat block. The blocks that are connected to a hat block will be executed at the time the condition on the hat block is met. We will be using this block a lot to create our first game.

Project 1: Interactive Art Project

This Scratch project (Figure 3-29) generates circles, squares, and triangles with different sizes and colors when a conductive keyboard object is touched. Start by double-clicking the Scratch icon on the desktop of the Raspberry Pi (Figure 3-28).

Figure 3-28. *Scratch shortcut*

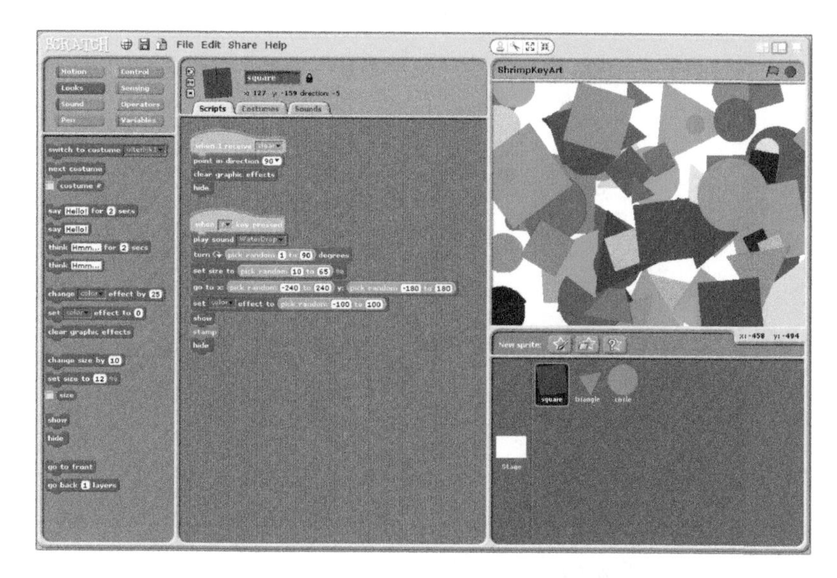

Figure 3-29. *Interactive art project*

1. Add three new sprites: a square, a triangle, and a circle. Making a new sprite in Scratch is done by clicking the "Paint new sprite" button (Figure 3-30).

   Figure 3-30. *New sprite button*

2. The Paint Editor opens (Figure 3-31).

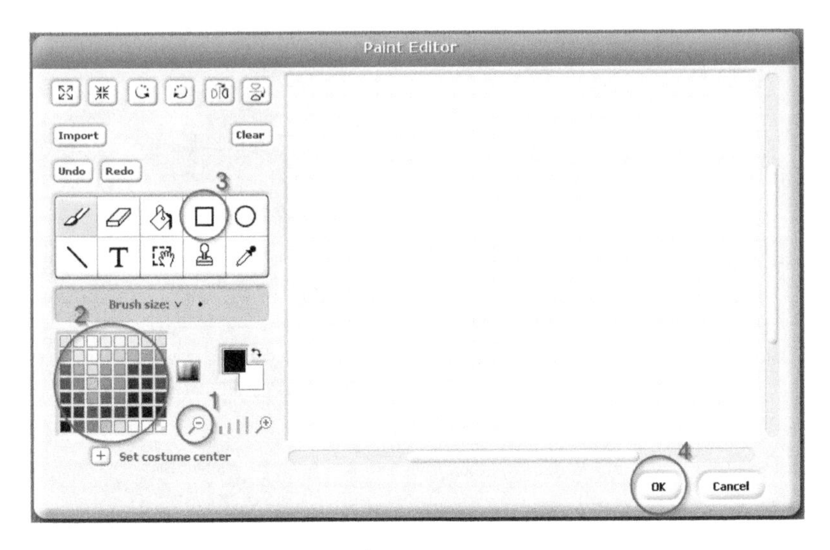

Figure 3-31. *Scratch Paint Editor*

3. Execute the following steps to make the first sprite:
   a. Click the "Zoom out" button in the editor screen.
   b. Pick a color.
   c. Select the Rectangle Tool and draw a square.
   d. By holding the Shift key you'll get a perfect square.
4. Click OK to exit the Paint Editor.

Click again on the "Paint new sprite" button and repeat these steps to draw a triangle with the Line Tool. Repeat the steps again to draw a circle with the Ellipse Tool.

You now have three sprites and the default Scratch Cat sprite. Let's remove the Scratch Cat in this project. Right-click the Scratch Cat, and select Delete from the menu.

Click the square sprite. At the top of the middle section of the screen you'll see Figure 3-32.

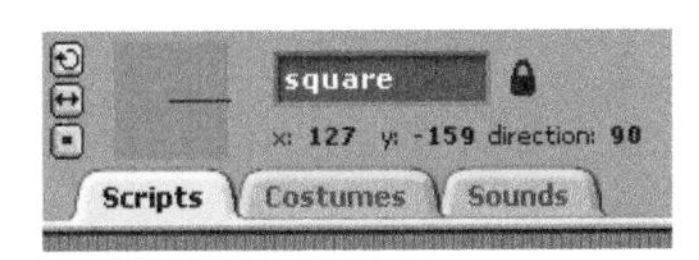

Figure 3-32. *Scripts, Costumes, and Sounds tabs*

You can change the name of your sprite here. You'll also see three tabs: Scripts, Costumes, and Sounds. Let's add a sound to this sprite so that it will play this sound whenever a square is drawn on the screen:

1. Click the Sounds tab.
2. Click the Import button.
3. Double-click the *Effects* folder and choose the "WaterDrop" sound by double-clicking it.

Repeat these steps for the triangle sprite (choose the "Rattle" sound) and the circle sprite (choose the "Pop" sound).

The stage is a special sprite. It controls the background of your project, but it's also the perfect place to add code to control your project.

Coding in Scratch

From here on we'll be coding. First, you'll see the finished script (a stack, or stacks, of blocks), and after that I'll explain these blocks. Every sentence will contain one or more words in *italic*. These words refer to one of the blocks in the script.

Click the stage icon and add the scripts shown in Figure 3-33 by dragging the blocks into the script zone and snapping them together.

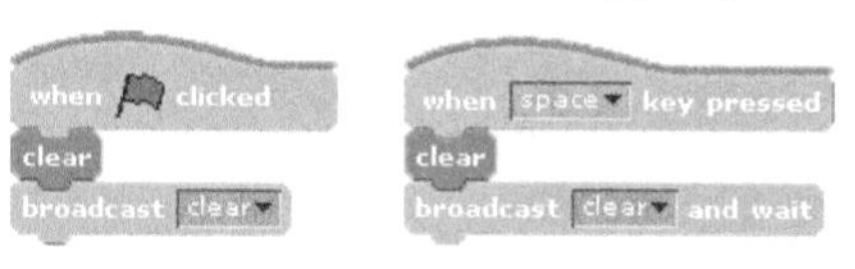

Figure 3-33. *Stage scripts*

The script on the left in the figure runs *when the green flag is clicked* and will *clear* the stage, and *broadcasts "clear"* to all of the sprites. The broadcast block sends a message to all of the sprites in a project that can trigger code in that sprite, which you will see later.

The script on the right in the figure runs *when space key pressed*, and will *clear* the stage. It then *broadcasts "clear" and waits* until the receiving scripts are finished executing.

Now click the square sprite. For the square sprite we're going to make two scripts. Code the script shown in Figure 3-34.

Figure 3-34. *First script for the square sprite*

Besides snapping blocks together under or above each other, it's also possible to snap blocks directly on top of each other:

This script is executed *when the s key is pressed* and will *play the sound WaterDrop*, then it will *turn* the sprite *at random between 1 and 90 degrees clockwise*. It also *sets the size at random between 10 and 65%* of the original size and will place the square *at random on an x: and y: coordinate*. It will then *set the color effect to a random color* and then it'll be *shown* on the stage. By *stamping* it, this square will be painted permanently on the stage, and then the sprite can be *hidden* again.

You also have to make a small script to clear everything when the project runs for the first time, or when the space bar is pressed (as coded on the stage; see Figure 3-35).

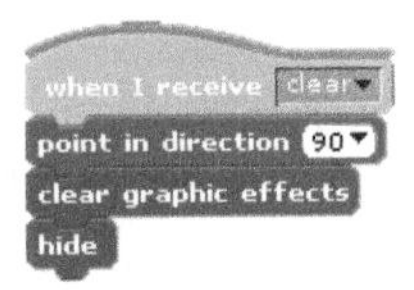

Figure 3-35. *Second script for the square sprite*

This script is executed *when it receives "clear"* (from the script on the stage) and will *point* the sprite *to the right (90 degrees)* (this will level your square) and *clear all graphic effects*, and then will *hide* the sprite.

Your Scratch window should look like Figure 3-36 now.

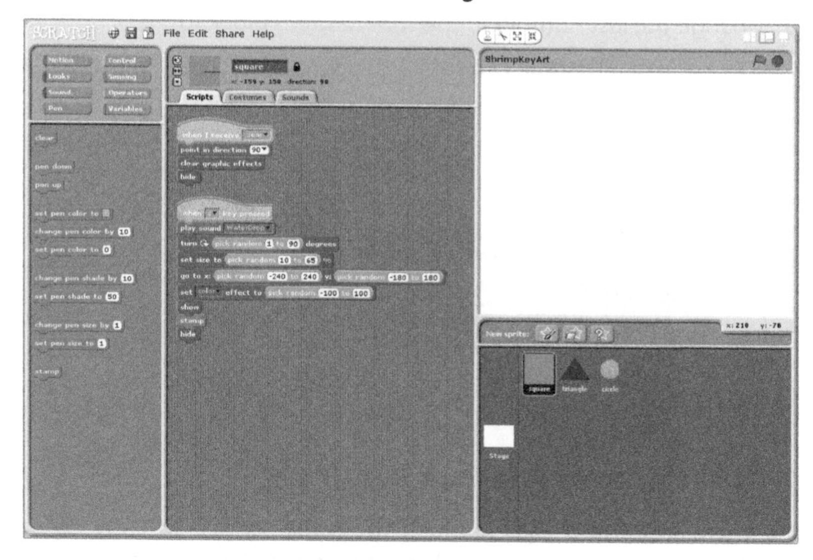

Figure 3-36. *Current state of Scratch window*

The code for the triangle is almost the same, so you can copy this to the triangle sprite and then make some changes.

Copying code to another sprite is easy:

1. Click the hat block of the script and (while holding the left mouse button) drag the script on top of the triangle sprite (a light gray box appears).
2. Release the mouse button now.

   The script will move back to it's original position, but it's also copied to the triangle sprite.
3. Repeat this for the other script.
4. Now click the triangle sprite.

If everything went well, you will have the same scripts in the triangle sprite. If not, repeat the steps for the script(s) that haven't been copied.

We're going to make some changes (see Figure 3-37):

1. Change the key in the *when key pressed* block to the letter *d*.
2. Change the sound in the *play sound* block to *Rattle*.
3. Change the last number in the *turn degrees* block to *120*.

   (One side of a triangle can turn 120 degrees before it's in the starting position.)

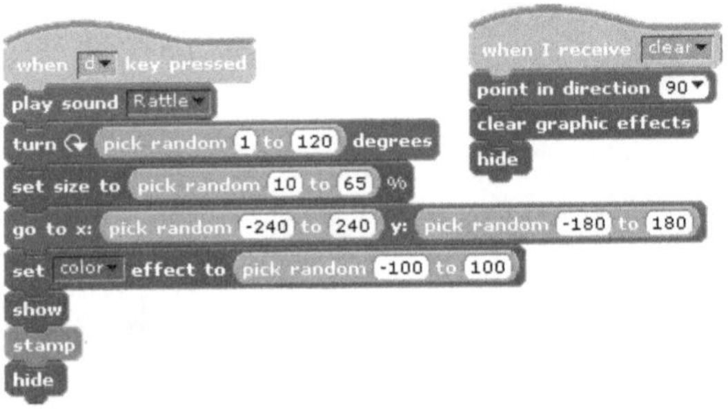

Figure 3-37. *Scripts for the triangle sprite*

The circle sprite also uses (almost) the same scripts, so copy them from the square sprite to the circle sprite. Now click the circle sprite.

We're going to make some changes to the circle:

1. Because turning a circle doesn't change its appearance, we're going to remove these blocks by using the scissor tool (it's above the stage).
   a. Click the scissor button and move it to the *point in direction* block.
   b. Click the block when the red box appears around it. This will remove the block.
2. Repeat this with the *turn degrees* block.
3. Change the key in the *when key pressed* block to the letter *a*.

The scripts for the circle should look like Figure 3-38.

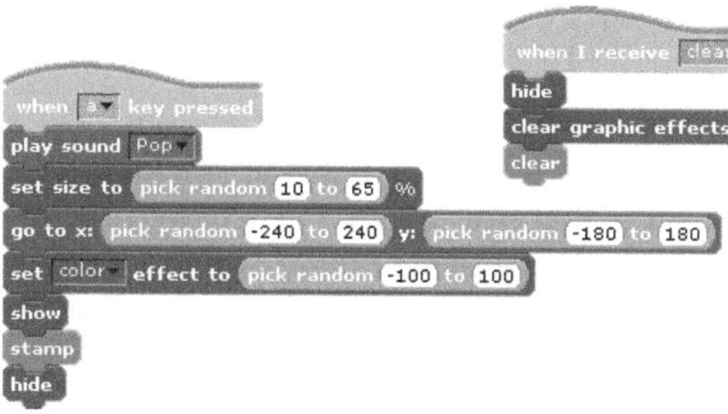

Figure 3-38. *Scripts for the circle sprite*

Your interactive art project is finished! Go to File → Save to save your project.

Now that the Scratch game is complete, you need to configure the conductive keyboard to interact with the game.

Using the Conductive Keyboard with the Art Project

We've used four keys in this Scratch project: *a*, *s*, *d*, and the *space bar*. On your conductive keyboard, these are mapped to pins *0*, *1*, *3*, and *9*.

Be creative and make a cool conductive keyboard for this project. You can use Play-Doh, fruit, vegetables, sauces, flowers, people, or any other conductive material. You need at least three objects to generate the randomly placed circles, squares, and triangles. The fourth object is to clear the stage. For players, it's best that you make a clear distinction between the different objects to explain how they are used.

The objects connected to pin *0* (a), *1* (s), and *3* (d) respectively draw a circle, square, and triangle. The object connected to pin *9* (space bar) clears the screen.

When you've finished your conductive keyboard, you have to connect the USB cable to the Raspberry Pi, and click the green flag in Scratch to run it.

Every time you touch an object your interactive art will change.

This project explained the basics of Scratch. The next project in this chapter is more advanced, since it uses the GPIO pins of the Raspberry Pi to turn on or off LEDs, to make a memory game.

Project 2: Memory Game

Many traditional memory games consist of four colored buttons. When playing the game, the device lights up one or more buttons in a random order. After that, the player must reproduce that order by pressing the buttons. As the game progresses, the number of buttons to be pressed increases, until you make a mistake in the reproduction of the order and have to start over.

For this project, we're going to build a memory game using Scratch as the "brains" (Figure 3-39). The lights will be LEDs connected to the Raspberry Pi's GPIO pins, and the buttons are made by you and are connected to the conductive keyboard (using *a*, *s*, *d*, and *w*).

Figure 3-39. *Memory game overview*

We're going to use a special version of Scratch that can control the GPIO pins: *ScratchGPIO* (Figure 3-40). It's open source and developed by Simon Walters (*http://simplesi.net*) and "is designed as an easy way to use Scratch on the Raspberry Pi to control lights/motors/sensors and switches using the GPIO pins."

Figure 3-40. *ScratchGPIO shortcut*

ScratchGPIO supports two kinds of functions: inputs and outputs. The output pins can be used to control LEDs or motors. The input pins can be used to detect button presses or to read sensor values. For this project we use the output pins. So, I'm only going to explain how to use these. For more information about other features and a command guide, visit Simplesi (*http://bit.ly/1sSSerj*). In Figure 3-41, you'll see the blocks that control the GPIO output pins.

Figure 3-41. *Blocks to control the GPIO output pins*

Turning GPIO pins on or off (or setting them high or low) can be done in a couple of ways. In this Scratch project we're going to use broadcast messages for this. The top two blocks set the power (+3.3V) on GPIO pin 11. The following two blocks unpower GPIO pin 11. The bottom two blocks set the power (*allon*) to all GPIO output pins or unpower (*alloff*) all of the GPIO output pins.

By default, the following GPIO pins are used as outputs: 11, 12, 13, 15, 16, and 18. We're using pins 11 (GPIO 17), 12 (GPIO 18), 13 (GPIO 27), and 15 (GPIO 22) for this project. Please look up these pins in Figure 3-42.

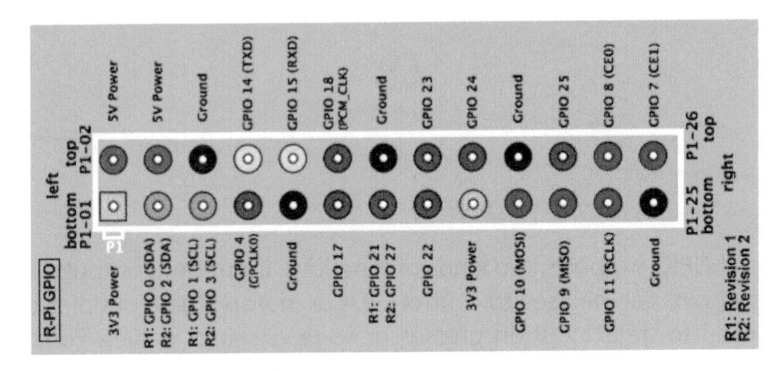

Figure 3-42. *GPIO pin numbering for Raspberry Pi B*

The Raspberry Pi uses +3.3V internally, but can output +5V. When connecting a +5V (and even a +3.3V) pin with another GPIO pin, you can damage this pin, so be careful.

Let's start making the Scratch project. Open ScratchGPIO5 by double-clicking its icon on the Raspbery Pi desktop. It will open Scratch. It looks like the normal version of Scratch (and it is). The only difference you'll notice is that upon opening, you'll get a pop-up about remote sensors being enabled (click OK to close it). In the background, a helper app also starts.

We'll start by making a list. A list in Scratch is a special variable. Where a variable can only store one value at a time, a list can store multiple values. It's also possible to change the order in a list, or to add or replace a value at a specific place in the list. We're using the list to store the sequence that has to be reproduced for our memory game.

Go to the Variables palette and click *Make a list*. Type **sequence** in the pop-up and click OK (Figure 3-43).

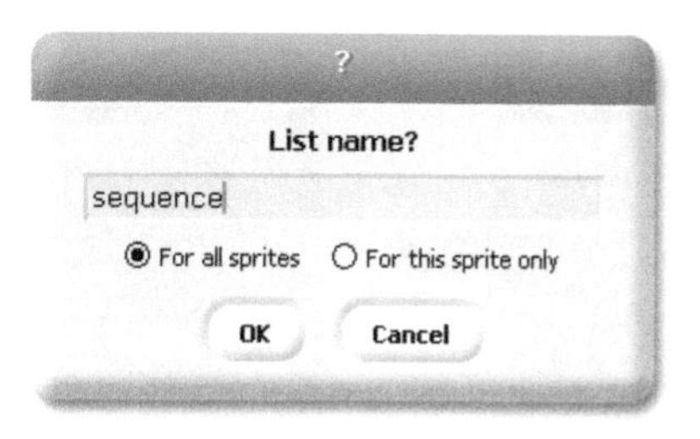

Figure 3-43. *Create a list*

This will add extra blocks on the Variables palette, which we need in our project. We also need more variables to store other values. Click *Make a variable* and create the following variables (one at a time):

check
: This variable is used to only detect key presses (by the conductive keyboard) when the played sequence has to be reproduced, and not while playing the sequence.

key
: This variable is used to store the pressed key and to check it against the value in the list.

number
: This variable is used to store the place in the list that has to be played (turning on the LED), or has to be checked (against the pressed key).

This action also adds extra blocks to the Variables palette.

We're using the Scratch Cat sprite as our "game master," so it can tell us how we're doing during game play. Add the scripts shown in Figure 3-44 to its sprite. After making the first one, right-click the hat clock and choose "duplicate" from the menu to make a copy. Then change the values in the *if key pressed*, *set key to*, and *broadcast* blocks.

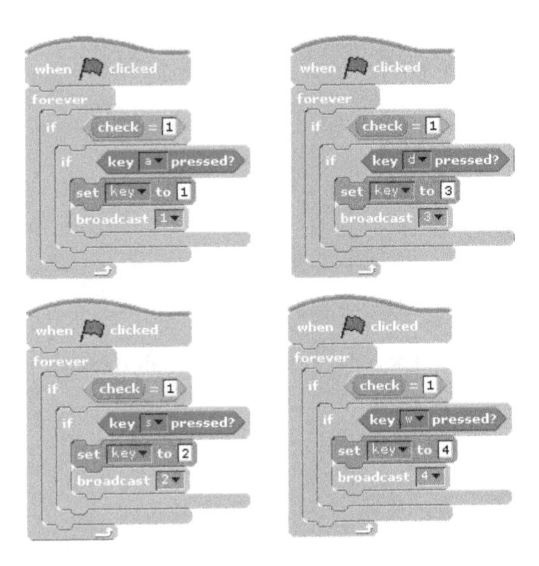

Figure 3-44. *First Scratch Cat scripts*

The blocks in Figure 3-44 are for detecting the pressed key (by your conductive keyboard), which is stored in the *key* variable. The script will only check for pressed keys if the variable *check* is set to *1*. As you can see, we're using value *1* for the *a*, *2* for the *s*, *3* for the *d*, and *4* for the *w*. Don't forget to place the *forever* block around the other blocks, otherwise it will only perform the check once when pressing the green flag.

Make the scripts in Figure 3-45 also.

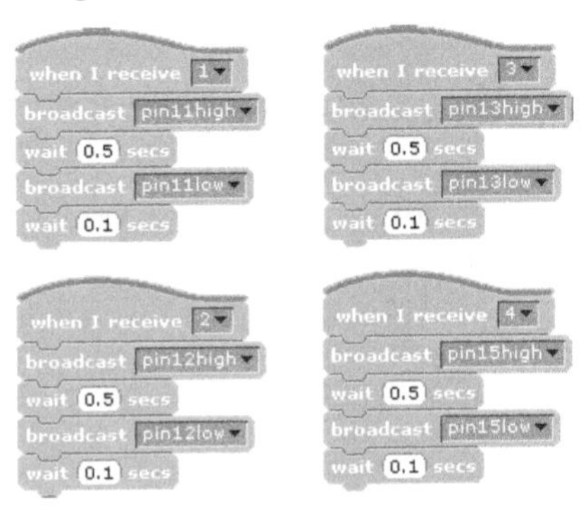

Figure 3-45. *Second Scratch Cat scripts*

These scripts will turn on the LEDs for our memory game. As you can see in Figure 3-44, we're using the *broadcast* blocks to send values 1–4. The hat blocks in these scripts continuously listen for broadcasts. When a script receives the right broadcast it will execute the blocks underneath it. In this case it will set one of the GPIO pins to high, wait 0.5 seconds, and then unpower the GPIO pin. To give it some time to do that, we'll be waiting 0.1 seconds for that to happen.

Now make the two scripts shown in Figure 3-46.

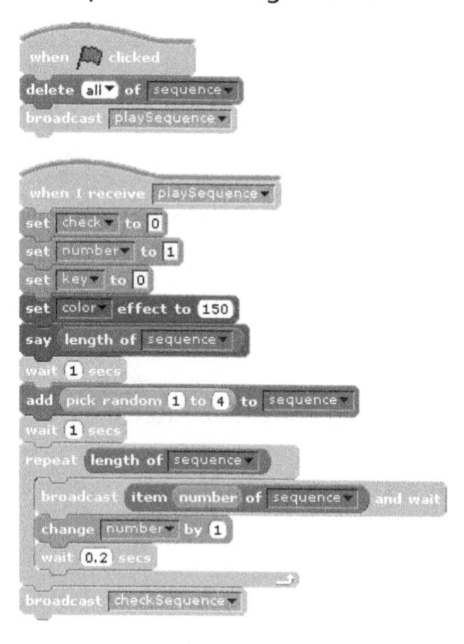

Figure 3-46. *Third Scratch Cat scripts (from elinux.org/File:GPIOs.png)*

The first script is executed when you click the green flag and clears the list (we'll always want to start with a clean list), and then broadcasts the message *playSequence*. This message is picked up by the second script in Figure 3-46.

If you're wondering why I didn't connect the second script under the first script in Figure 3-46, it's because by using the *when I receive* hat block you can reuse the script later on. After playing and checking the first value, we'll go back to this script to add the second value, and so on. Without this hat block, we would have to add these blocks a second time to our project, and a third time, and a fourth, and so on. That would be very laborious.

Every time this script is executed it *sets check to 0* to disable the key checking. It also *sets number to 1* because we are starting at place 1 in the list (since the sequence is played from the beginning every time). Lastly, it *sets key to 0* to delete the last pressed key.

We now *set the color effect to reddish* (the Scratch Cat will turn red to signal that it doesn't check for key presses) and the Scratch Cat will *say* (in a text balloon) *the length of the sequence* (this tells you how many runs you already did correctly). After that it *waits for 1 second* and then it will *add a number from 1 to 4 to the sequence list*.

After *waiting another second* it will *broadcast the value on the first (or second, and so on) place of the list and waits* until the script that turns on and off the LED is finished. It will *change the number by 1* to go to the next value on the list, and *waits for 0.2 seconds*. These three blocks will be *repeated* the same number of times as the number of values in the list (*length of sequence*).

When all of the values are played it will *broadcast checkSequence*, and this will trigger the script in Figure 3-47 that you'll have to code.

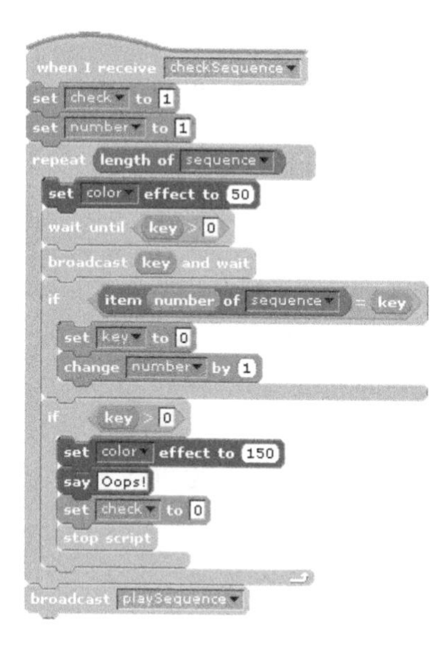

Figure 3-47. *Fourth Scratch Cat script*

The script *sets the check variable to 1* to turn on the checking mechanism (the first scripts you've made). It also *sets number to 1* since we have to start at place 1 in the list again (but now we have to check it against the pressed key).

The following blocks will be *repeated* the same number of times as the number of the values in the list (*length of sequence*):

set color effect to 50
: Will give the Scratch Cat a greenish color.

wait until key > 0
: We now have to wait until a key is pressed, by checking the *key* variable for a number larger than 0 (this is set by the first scripts you made).

broadcast key and wait
: Will turn on the LED that has the same value as the pressed key.

if item number of sequence = key
: Now it checks the pressed key against the place in the list we want to check. If it matches, it will execute the following blocks.

set key to 0

We reset the pressed key, otherwise it will also check this value against the next item in the list.

change number by 1

We'll move down a place in the list.

if key > 0

Will only be executed if the previous condition hasn't been met (because we're resetting the *key* variable to 0 when the right key is pressed), which means that the wrong key has been pressed.

set color effect to 150

The Scratch Cat will turn red again.

say Oops!

The text in the text balloon will be changed to "Oops!"

set check to 0

Turns off the checking mechanism.

stop script

Stops this script because you've lost this game (restart it by click the green flag).

If everything went well (the correct sequence has been reproduced) it will *broadcast playSequence* and the next (and longer) sequence will be played (see the previous script).

We've now finished building the Scratch project, which supplies us with the "brains" of our game. Let's continue building the hardware.

We need:

- 4x LEDs (preferably different colors)
- 4x 330 Ω resistors
- 10x jumper wires
- 1x breadboard

Place the four LEDs and the 330 Ω resistors on the breadboard as shown Figure 3-48 and the following chart.

Figure 3-48. *Memory game schematics*

Wiring on the Raspberry Pi B+

The wiring for a Raspberry Pi B+ is the same as for the Raspberry Pi B, even though the B+ has 12 more pins. Just connect the wires as shown in Figure 3-48. The 12 extra pins on the B+ are at the right of the black wire and aren't used in this section.

Component	From	To
LED	short leg B:9	long leg C:6
LED	short leg C:9	long leg D:12
LED	short leg G:9	long leg H:6
LED	short leg H:9	long leg I:12
330 Ω resistors	B:1	B:6
330 Ω resistors	C:12	C:17
330 Ω resistors	G:1	G:6
330 Ω resistors	H:12	H:17

Connect four jumper wires to pins 11 (GPIO 17), 12 (GPIO 18), 13 (GPIO 27), and 15 (GPIO 22). To be absolutely safe, do this while the Raspberry Pi is turned off, and connect them to the breadboard as shown in Figure 3-48.

Also, attach a wire to the GND GPIO pin and the breadboard.

Component	From	To
wire	RPi GPIO17	F:1
wire	RPi GPIO18	A:1
wire	RPi GPIO 27	A:17
wire	RPi GPIO 22	F:17
wire	RPi GND	A:9
wire	E:9	F:9

RasPiO

If you don't feel comfortable connecting wires directly into your Raspberry Pi, you could buy an add-on board. One of them is the RasPiO (*http://rasp.io/breakoutpro/*). The RasPiO has 330 Ω resistors connected to each pin, which would allow you to skip those on the breadboard.

You've now finished the hardware portion of this memory game and you can start connecting your conductive keyboard.

Make four keys for the conductive keyboard (be creative) and attach them to pins 0 (*a*), 1 (*s*), 3 (*d*), and 4 (*w*) of the keyboard. Now attach the USB cable to the Raspberry Pi. Make sure the four keys are placed in the same order as the LEDs, or place them around the breadboard, one at each corner.

Component	From	To
wire	ShrimpKey pin 0	object
wire	ShrimpKey pin 1	object
wire	ShrimpKey pin 3	object
wire	ShrimpKey pin 4	object

Now click the green flag in Scratch and watch the first LED light up. Remember it and press the corresponding key when the Scratch Cat turns green. Watch the LEDs again. Next, it will turn on two LEDs, one after the other. Reproduce that sequence, and so on. Good luck!

Summary

Earlier in this chapter you created a conductive keyboard with components (including the ATmega chip) and breadboards, using the Raspberry Pi. You then made an interactive art project and a memory game in Scratch, both controlled by the conductive keyboard that you built. For the memory game, you also used the GPIO pins to control the necessary LEDs. I hope you enjoyed completing these projects, and that you can continue making and transforming your own ideas into exciting Maker projects.

4 The Internet of Fish

By Lauren Orsini

Fish make low-maintenance pets, but they don't communicate well. An unhappy dog, cat, or even a parrot will let you know if something's wrong. With fish, however, it's often hard to know if they're feeling OK until it's too late. Fortunately, technology can help by monitoring the environment of your fish (see Figure 4-1).

One thing fish are especially sensitive to is water temperature. A temperature fluctuation of 10 degrees in either direction can shock and kill a fish. That's where the Raspberry Pi comes in.

Sitting next to my aquarium, the Raspberry Pi is performing three tasks:

1. The Pi uses a waterproof thermometer output to read the aquarium's temperature. This is no ordinary thermometer, but a "smart" one that is capable of transferring data. Using a solderless breadboard, you will build a circuit that connects the thermometer to the Raspberry Pi, so it can transfer data.
2. The Pi texts your cell phone in overheating or cooling emergencies. The data coming from the thermometer serves as an API, an application programming interface, that other programs can acquire data from. In this case, you'll use an SMS service called Twilio to assign a phone number to the Pi so you can receive texts.

3. The Pi is powering a local web server that reads and visualizes the thermometer's data. Even when the water temperature isn't in a danger zone, you can type in the Pi's local address (or, if you want to make it public, a URL). You can then monitor rises and falls in water temperature whenever you're curious. The Pi hosts a local website that displays a line graph, showing you how the aquarium's temperature has been trending.

This project is essentially a smart, communicative thermometer—the Internet of Things for fish!

Figure 4-1. *Internet of Fish project*

Of course, it helps to have a smart thermometer for a number of different reasons, not just indoor aquariums. You could monitor an outdoor pond by modifying the program to account for the wider temperature fluctuations that occur outside. Maintaining a perfect temperature is also essential for an at-home brewery, and this could be a technology-forward way to help assure that.

Materials

The required materials are divided into hardware and software sections.

Hardware Materials

First, let's review the hardware (listed in Table 4-1 and shown in Figure 4-2).

Table 4-1. *Bill of materials*

Raspberry Pi (Models B or B+)
Pi Cobbler
Breadboard
4.7k (1/4 watt) resistor
Pack of M-M jumper wires
Wireless USB adapter
DS18B20 digital temperature sensor

Raspberry Pi and everything that goes with it
: This is a package of components including an SD card, a power source, and a case to put the Pi in (it isn't a good idea to put an uncased Pi next to a fish tank).

Breadboard
: In this case, you're using a solderless breadboard. A miniature breadboard with 400 connection points works well for this project.

Pack of male to male jumper wires
: You'll want a good selection of colors and lengths so you can tell the connections apart. This project uses six wires, and I found it helpful to use three different colors—red, orange, and black—to show the three different connections.

DS18B20 waterproof digital temperature sensor
: Adafruit is a great resource for Raspberry Pi. Depending upon the version, you'll see either three or four wires poking out of the bottom of the sensor. The DS18B20 comes in both regular and high-

temperature-resistant versions, but either one will work for this project.

If the DS18B20 waterproof digital temperature sensor is a three-wire model, the wires will be red, white, and black. Red is for the power connection, white is for the data connection, and black is for grounding the circuit. If it's a four-wire model, they'll be red, white, black, and orange. In this version, red is still power and black is still ground, but it's orange that is the data connection wire. In the four-wire model, you leave the white wire disconnected.

Pi Cobbler
: This 26-pin header is an Adafruit creation that takes all of the useful input/output pins on top of the Pi and transfers them, via GPIO cable, over to your breadboard. The Cobbler keeps your Pi protected in a case. The Pi Cobbler usually comes with a GPIO ribbon, but if you get it second-hand, you'll have to purchase one separately.

4.7k (1/4 watt) resistor
: The temperature sensor requires a very low voltage, even lower than the one the Raspberry Pi circuit naturally conducts. A 4.7k ohm resistor reduces the amount of current that goes through the sensor to the degree recommended by the DS18B20 manufacturers. It helps to have a packet of resistors at the ready, since they're tiny and fragile and can be lost or broken.

Wireless USB adapter
: This tiny device gives the Pi the ability to connect to Internet WiFi via one of its USB outlets. When you have a WiFi connection, you're able to operate the Pi "headlessly," that is, without a keyboard, mouse, or monitor. For the majority of this project, we'll be connecting to the Pi over a local wireless connection. It is a good idea to review the compatible hardware list (*http://elinux.org/RPi_USB_Wi-Fi_Adapters*) of USB adapters that are known to work with the Pi.

Figure 4-2. *All the hardware you'll need*

Software Materials

All of the software used for this project is free, and nearly all of it is open source. I prefer to use open source software for development. I like the idea that an entire community of developers always "has my back," since somebody's always working to improve and offer advice about it.

Internet of Fish Source Code

Be sure to download the source code for this project. It is located on GitHub (*https://github.com/backstopmedia/maker/*), but see Appendix A for instructions on downloading the Fishtank source code to the Rasberry Pi desktop.

NOOBS (New Out Of Box Software)
: You'll need to load this to the SD card when you boot up the Pi for the first time.

Twilio
: A developer-friendly set of tools for creating SMS, voice, and VoIP applications. You'll be using the Twilio API to assign a phone number to your Raspberry Pi so it can send texts. Twilio usually costs money to use, but it's free if you only use it to text yourself, like we're doing for this project.

MySQL
: Open source software for database management. Databases are where you organize and store collections of information in a way that's easy for a computer to read. This is where the project will collect and store sensor data from the waterproof thermometer.

python-mysqldb
: A Python interface for MySQL. Python, the Raspberry Pi's development language of choice, doesn't naturally work with MySQL. This lightweight interface will ensure that your Python program communicates with the database you build.

Screen
: A free Linux program that keeps programs running in the background even while you're working on other things. The goal is for the smart thermometer to eventually run constantly, and this will help you set that up.

The Apache HTTP Server (Apache for short)
: A web server application that you'll use to set up the Pi so it can host its own website.

D3.js
: An open source program for building lightweight, pretty graphs. Even better, it uses JavaScript. You'll use this to make the display graph on the Raspberry Pi web server.

Before You Start

Let's assume you've just taken your Raspberry Pi out of the box. In that case, there are two things you need to accomplish before starting this project. First, you need to install Raspbian (see Appendix A). Second,

you need to set up a Secure Shell connection (SSH), which we'll do next.

Setting Up SSH Communication

It isn't a good idea to set up a Raspberry Pi, complete with monitor, mouse, and keyboard, right next to your fish tank (or any source of water). You can protect the Raspberry Pi by housing it in a waterproof case, but this leaves the rest of the electronic components vulnerable. To get around this, we'll be wirelessly transmitting our data. To keep the data secure, we'll be setting up an SSH connection.

With SSH you can operate the Pi "headlessly," without it being physically connected to a keyboard, mouse, or monitor. As long as you're still on the same wireless connection, you don't even have to be in the same room—or house—as your Pi.

SSH now comes preinstalled in Raspbian, the latest version of the Raspberry Pi operating system. Fortunately, that's the OS we just installed (shown in Appendix A)!

To use SSH, you need your Pi's IP address. Let's get that. Without the wireless USB adapter plugged in, launch the LXTerminal in Raspbian, and type the following into the Raspberry Pi's command line:

```
sudo nano /etc/network/interfaces
```

Nano is a default text editor on the Raspberry Pi that we'll be using a lot in this project to edit files. Adjust the contents of the file until it looks like the following code. Of course, make sure that "ssid" and "password" are replaced with your actual network name and password!

```
auto lo

iface lo inet loopback
iface eth0 inet dhcp

allow-hotplug wlan0
auto wlan0


iface wlan0 inet dhcp
        wpa-ssid "ssid"
        wpa-psk "password"
```

Now, shut the Pi down with the following command:

```
sudo shutdown now
```

Then, plug the wireless USB adapter in. Reboot your Pi to the command line and type:

```
sudo ifconfig
```

In response to this command, the Pi should spit out three paragraphs. Your IP address will show up in either the first or the third line, depending on whether your Raspberry Pi is hooked up to an Ethernet cable or via a WiFi adapter. Since we're using WiFi, look in the line that begins with `wlan0`.

You'll see the words `inet addr` followed by an IP. That will be something like 192.168.2.2, a pretty common default IP address that I'll use as the example address for now.

This address allows you to access the Pi from your computer. If you're on a Mac, you already have built-in SSH. Launch the Terminal application from your Mac OS and type:

```
ssh pi@192.168.2.2
```

It'll ask for your password. By default, this is always "raspberry." If you've changed it to something else, use that instead. Now, you're in!

If you're on a PC, there's an extra step.

Download and run PuTTY or another SSH client for Windows. Enter your IP address in the field, as shown in Figure 4-3. Keep the default port at 22. Press Enter, and PuTTY will open a terminal window that will prompt you for your username and password.

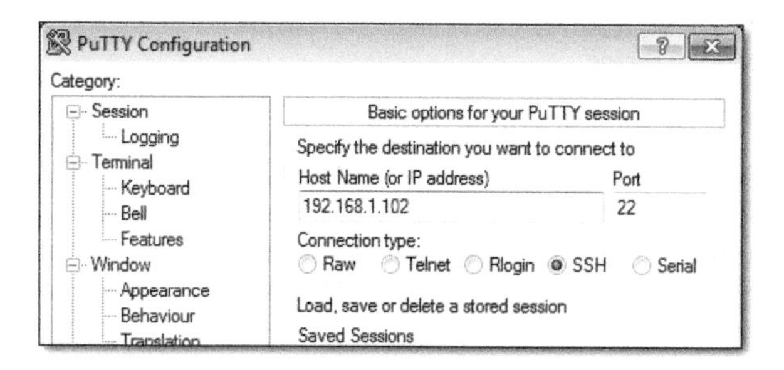

Figure 4-3. *Running PuTTY on a Windows machine*

Fill those in, and you're all set. You may now begin working remotely on your Pi.

Figure 4-4 is a screenshot of me connecting to my Pi from my Macbook computer. Even though I'm not anywhere near my Pi, I'm working on it!

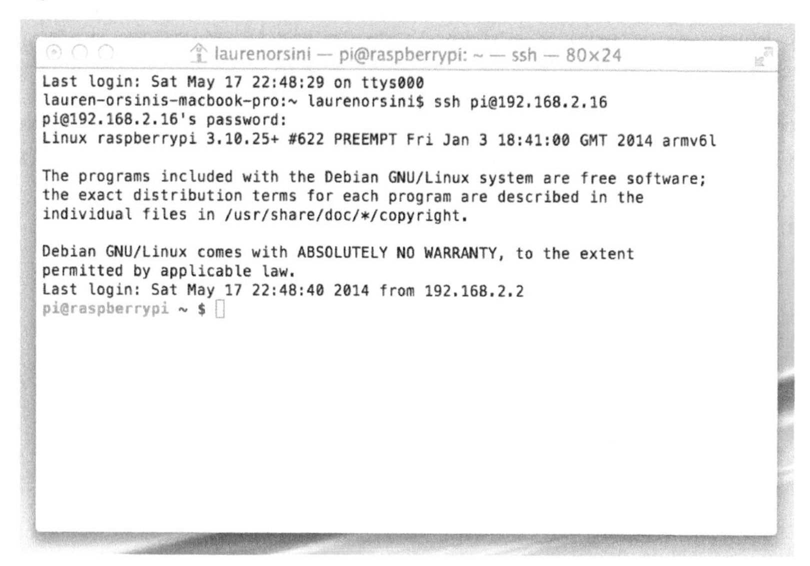

Figure 4-4. *A working SSH connection between an OS X computer and a Raspberry Pi*

Now that we're all set up, go back to your Raspberry Pi. Type the following from the command line:

```
sudo apt-get update
sudo apt-get upgrade
```

This first command will inform the Pi that new changes are available and the second command will actually upgrade the entire system. This is going to save you so much trouble later on. Nearly every issue I've ever had with installations has been simply because I hadn't updated recently enough. When in doubt, update!

A Thermometer That Reads and Writes

The heart of the hardware side of this project is the DS18B20 waterproof thermometer. It's the most complicated part (aside from the Pi!) since it converts real sensory data into computer readings that the Pi can understand.

As sensors go, the DS18B20 is a smart one. It records the temperature in Kelvin down to several decimal places, but I prefer Fahrenheit or Celsius.

It's time to turn off the Raspberry Pi. It's a bad idea to connect live wires! It can be dangerous to unplug a Pi without properly shutting it down, so any time you need to turn this off, type this into the command line first:

```
sudo shutdown now
```

Now, gather your materials. For this section you'll need:

- Breadboard
- Male to male jumper wires
- 4.7k resistor
- Pi Cobbler
- GPIO ribbon cable

You're going to be connecting the wires into a circuit as shown in Figure 4-5.

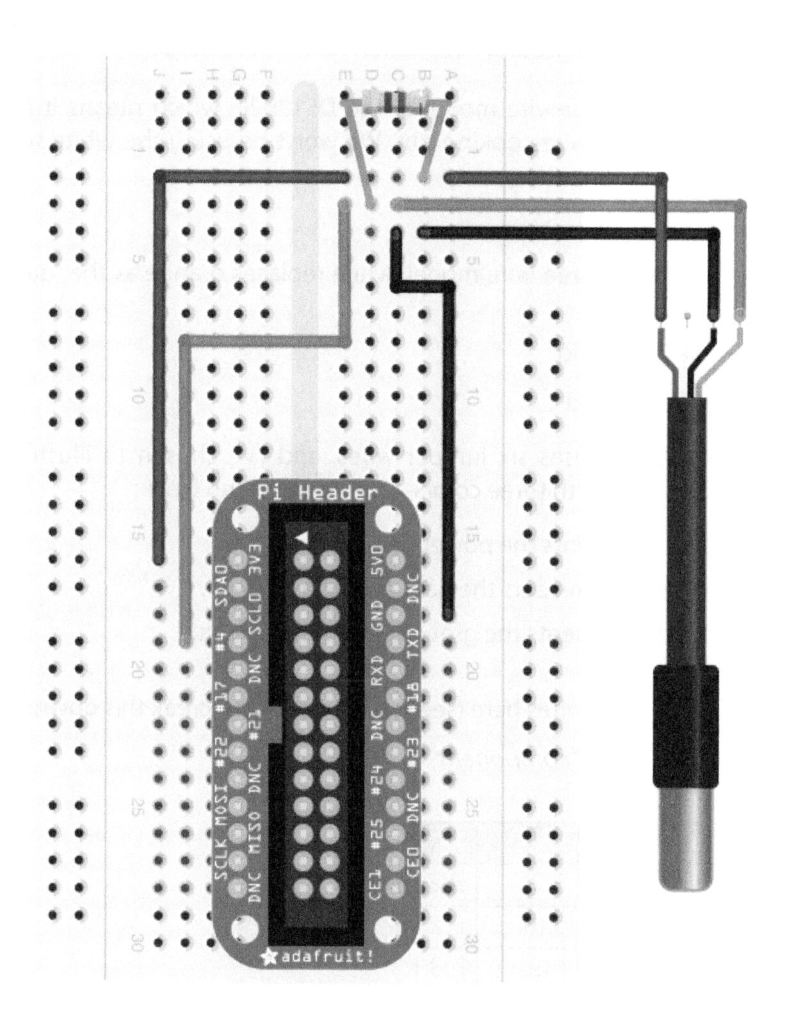

Figure 4-5. *The circuit diagram*

There are a few things to note here:

1. This electrical circuit is accomplishing three things:

    • Establishing a data connection between the thermometer and the Pi

    • Establishing a power connection from the Pi to the thermometer

- Setting up a ground connection for return voltage

2. This is the four-wire model of the DS18B20, which means it has four colored wires poking out. We won't be using the white wire as previously mentioned.

   - Red: power
   - White: in three-wire model, white replaces orange as the "data" wire
   - Black: ground
   - Orange: data

3. This project uses six jumper wires, and I've chosen to illustrate those wires with three colors—red, orange, and black:

   - Red represents the power part of the circuit.
   - Orange represents the data part of the circuit.
   - Black represents the ground part of the circuit.

There are a lot of wires here (see Table 4-2), so let's break this down.

Table 4-2. *Table of component placement*

Component	From	To
4.7 kΩ resistor	D:3	B:2
wire 3v3	J:16	E:2
wire GPIO#4	I:19	E:3
wire GND	A:18	C:4
wire DS18B20 1	power	A:2
wire DS18B20 2	data	C:3
wire DS18B20 3	ground	B:4

The red wire section supplies power from the Pi to the thermometer. In this case, it's delivering a 3.3V connection, the lowest the Pi will transfer. Even that is still too strong, so in between the red wire connected to the Pi Cobbler and the one connected to the DS18B20, we've placed a 4.7k resistor to keep the current from overpowering the thermometer.

The black wire section is the ground connection. It connects GND (ground) on the Cobbler to the 4.7k resistor. A ground connection is the return path for the circuit, and is required by every electrical circuit.

The data connection—the orange wire—is the most important part of the circuit! You are using General-Purpose Input/Output (GPIO) Pin #4 for the purposes of the project. It connects the data wire on the thermometer to Pin #4, which delivers the information values to the Pi.

Secure the red, orange, and black jumper wires to the respectively colored wires poking out of the connection end of the DS18B20. It is extremely important that the wires connect, and if you are having issues getting the thermometer running, I assure you that 9 times out of 10, it is because your wires are not properly connected. There are several methods you can use to connect wires:

Tape
: I used masking tape because it's clear and I could see at a glance if the connection had broken. In my first time around, I even used painter's tape, so really any adhesive will do!

Crimping
: You can crimp the two wire ends together with a crimping tool or even a simple wire cutter, using its blunt side.

Soldering
: If you have a soldering iron handy, this is probably the most reliable method. Soldering fuses the two wires together so their connection is secure.

Connect the breadboard circuit with the powered-down Raspberry Pi. You connect the GPIO ribbon cable by snapping it into place on top of the Pi Cobbler. At this point, you can even put the thermometer in water if you like, but make sure you always ensure a drip loop to avoid water damage. The thermometer may be waterproof, but your circuit and Raspberry Pi are definitely not!

Now turn the Pi back on using an SSH connection. I'm using the IP address we established earlier, but replace it with your own address:

```
ssh pi@192.168.2.2
```

It's time to make sure the DS18B20 is properly connected. If it isn't, you'll know right away since these commands won't return anything. If the DS18B20 doesn't return anything, it means you need to go back. Make sure your circuit matches the diagram, and make sure the wires are secure.

This series of commands on your Raspberry Pi requires you to go into super user (sudo) mode and attempt to access the thermometer. You're going to change directories (cd) to access the directory that it's in, and list (ls) the files that are inside the directory. The file containing the thermometer data should be inside the device's directory.

```
sudo modprobe w1-gpio
sudo modprobe w1-therm
cd /sys/bus/w1/devices/
ls
```

If all goes according to plan, a serial number unique to your DS18B20 will pop up. So type:

```
cd 28-xxxx
cat w1_slave
```

Change the xxxx to match the serial number that pops up.

The Pi should spit back some numbers that look like this:

```
7a 01 4b 46 7f ff 06 10 0b : crc=0b YES
7a 01 4b 46 7f ff 06 10 0b t=23625
```

Figure 4-6 is an actual image of what popped up on my screen.

```
laurenorsini — pi@raspberrypi: /sys/bus/w1/devices/28-000005458...
Desktop  ocr_pi.png  python_games
pi@raspberrypi ~ $ sudo modprobe w1-gpio
pi@raspberrypi ~ $ sudo modprobe w1-therm
pi@raspberrypi ~ $ cd /sys/bus/w1/devices/
pi@raspberrypi /sys/bus/w1/devices $ ls
28-000005458143  w1_bus_master1
pi@raspberrypi /sys/bus/w1/devices $ cd 28-000005458143
pi@raspberrypi /sys/bus/w1/devices/28-000005458143 $ cat w1_slave
7a 01 4b 46 7f ff 06 10 0b : crc=0b YES
7a 01 4b 46 7f ff 06 10 0b t=23625
pi@raspberrypi /sys/bus/w1/devices/28-000005458143 $
```

Figure 4-6. *A correctly installed DS18B20 providing feedback on the command line*

You can already tell this means there is temperature data being processed.

Now, let's write a program to make this temperature data more readable. Let's call the program *temp.py*, and build it in the Pi folder.

So type:

```
cd ..
```

Once you're at the top directory, type:

```
cd /home/pi
sudo nano temp.py
```

This will open up the */pi* directory and create a blank file by the name specified in the Nano default text editor.

The Code

The base code for this project comes from Simon Monk (*https://learn.adafruit.com/users/16*) of Adafruit (*https://adafruit.com*). Simon gave me permission to repost his program in this book. Thank you, Simon! Over the course of this project, we will be heavily modifying Simon's code.

First, we're importing three Python language libraries. Libraries help us shortcut commands that Python doesn't already recognize by default:

```
import os
import glob
import time
```

Next, we need to define a few variables in the glob library that tell the program where we're looking for the thermometer and its data:

```
base_dir = '/sys/bus/w1/devices/'
device_folder = glob.glob(base_dir + '28*')[0]
device_file = device_folder + '/w1_slave'
```

Let's define `read_temp_raw`, a variable that tells the program how to read the temperature in its original Kelvin off of the DS18B20.

We're creating a variable called "f," which stands for "file," since we want to open up the DS18B20's device file and make it readable. Then, we create an object from that data `f.readlines`, and use the `readlines` function of that object to return the value stored inside that file —the raw temperature data—before closing the file. The program then continues, returning a value called `lines`, which contains the object we read from the file.

In less technical terms we're:

- Opening a file
- Making the file readable
- Reading the file
- Storing what was read as an object
- Closing the file
- Ending up with a value called `lines`

```
def read_temp_raw():
    f = open(device_file, 'r')
    lines = f.readlines()
    f.close()
    return lines
```

Now, we need to make the `lines` value a temperature value that we can actually read. Depending on where you are in the world, that's either Fahrenheit or Celsius. Here, the program is reading `lines` in Kelvin every millisecond, dividing it by 1,000 to get the Celsius reading, and then multiplying that by 9 divided by 5 plus 32 to get Fahrenheit:

```
def read_temp():
    lines = read_temp_raw()
    while lines[0].strip()[-3:] != 'YES':
        time.sleep(0.2)
        lines = read_temp_raw()
    equals_pos = lines[1].find('t=')
    if equals_pos != -1:
        temp_string = lines[1][equals_pos+2:]
        temp_c = float(temp_string) / 1000.0
        temp_f = temp_c * 9.0 / 5.0 + 32.0
        return temp_c, temp_f
```

If you're not interested in Fahrenheit, you can easily comment it out like I've shown here (Python will ignore anything with a pound sign in front of it):

```
        temp_c = float(temp_string) / 1000.0
        # temp_f = temp_c * 9.0 / 5.0 + 32.0
        return temp_c, # temp_f
```

Finally, we need to tell it to `print` the temperature, that is, deliver the temperature to your computer screen:

```
while True:
    print(read_temp())
    time.sleep(1)
```

This will print the temperature to the screen every single second.

So once again, here is the entire program:

```
import os
import glob
import time

os.system('modprobe w1-gpio')
os.system('modprobe w1-therm')

base_dir = '/sys/bus/w1/devices/'
device_folder = glob.glob(base_dir + '28*')[0]
device_file = device_folder + '/w1_slave'

def read_temp_raw():
    f = open(device_file, 'r')
    lines = f.readlines()
    f.close()
    return lines

def read_temp():
    lines = read_temp_raw()
```

```
    while lines[0].strip()[-3:] != 'YES':
        time.sleep(0.2)
        lines = read_temp_raw()
    equals_pos = lines[1].find('t=')
    if equals_pos != -1:
        temp_string = lines[1][equals_pos+2:]
        temp_c = float(temp_string) / 1000.0 # turns Kelvin into Celsius
        temp_f = temp_c * 9.0 / 5.0 + 32.0 # turns Celsius into Fahrenheit
        return temp_c, temp_f

while True:
    print(read_temp())
    time.sleep(1)
```

Paste this code into *temp.py* using the Nano editor. Press Ctrl-X to save and exit the file.

Now it's time to make sure the program is actually working. Run the program from the command line by typing:

```
sudo python temp.py
```

If it's all good, you'll see a series of temperature readings appear on the screen every second (Figure 4-7).

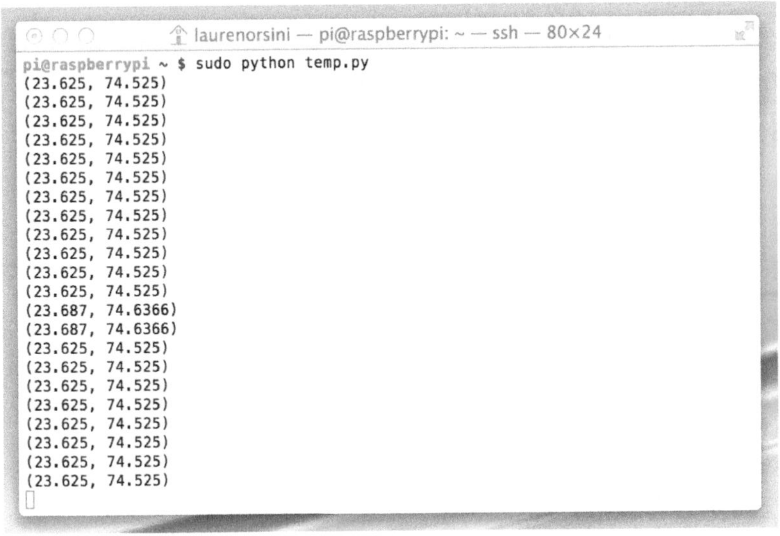

```
laurenorsini — pi@raspberrypi: ~ — ssh — 80×24
pi@raspberrypi ~ $ sudo python temp.py
(23.625, 74.525)
(23.625, 74.525)
(23.625, 74.525)
(23.625, 74.525)
(23.625, 74.525)
(23.625, 74.525)
(23.625, 74.525)
(23.625, 74.525)
(23.625, 74.525)
(23.625, 74.525)
(23.625, 74.525)
(23.625, 74.525)
(23.687, 74.6366)
(23.687, 74.6366)
(23.625, 74.525)
(23.625, 74.525)
(23.625, 74.525)
(23.625, 74.525)
(23.625, 74.525)
(23.625, 74.525)
(23.625, 74.525)
(23.625, 74.525)
```

Figure 4-7. *Temperature readings printing to the command line*

Try dunking the business end of the thermometer in ice water to cool it down, or breathe on it to warm it up. Watch the temperature fluctuate. It's working!

Now, any time you want to know what the fish tank's temperature is, connect to the Raspberry Pi with SSH and type:

```
sudo python temp.py
```

It will continue to display the temperature every second until you terminate the program.

You've now created a functioning smart thermometer that sends data to the Raspberry Pi. It's powerful enough to send new data every second. Next, you are going to put that smart thermometer to work.

Texting with Raspberry Pi

Once you get over the initial excitement, the limitations of the Pi-connected thermometer are pretty clear. The thermometer only gives you the temperature if you're on your home WiFi network, well within easy range of your Pi.

Amazingly, there is a way to free yourself from this limitation. The solution is Twilio, a developer-friendly set of tools for creating SMS, voice, and VoIP applications. Twilio charges pennies for calls and text messages to any phone, but it's free to develop programs that send texts to your own phone.

We're going to access the Twilio API to give the Python program additional capabilities—in this case, to send text messages.

Sign Up for Twilio

Go to *https://www.twilio.com* and set up your free Twilio account. Once you're up and running, Twilio will assign you a brand new phone number. It'll probably begin with the area code of whichever state or country you were in when you signed up. This is the phone number you will soon assign to the Raspberry Pi to give it SMS capabilities.

On your Twilio dashboard, you'll see one long number, and one long line of asterisks at the top. These are your SID (Security IDentifier) and your Auth (Authorization) Tokens. Without these two unique strings of numbers, you won't have permission to access the Twilio API. Copy these down somewhere so you can easily paste them into code when needed.

Figure 4-8 is a screenshot of what those API credentials look like, with the SID blurred out.

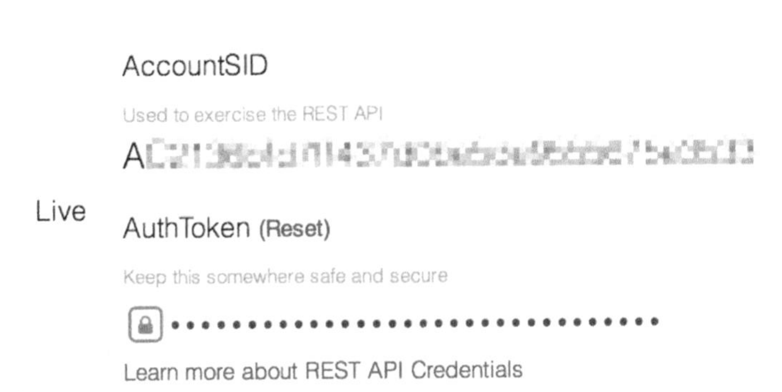

Figure 4-8. *API credentials*

Put Raspberry Pi in the Loop

You've set up a Twilio account so you can access the API, but Raspberry Pi doesn't know that yet. The Pi doesn't even know what Twilio is to begin with. So now you need to install the Twilio library on the Raspberry Pi.

Since Twilio is a Python interfacing library, you are going to use pip, Python's default package installer. Since the Pi comes with Python, pip might already be installed. To check, you can simply type `pip` into your command line while SSH'd to the Raspberry Pi.

If your Pi returns the message `-bash: pip: command not found`, that means pip still needs to be installed. You can do this on the command line with:

```
sudo apt-get install python-setuptools
sudo easy_install pip
```

Once you're certain pip is installed, you can use pip as a Twilio installer:

```
sudo pip install twilio
```

Make sure you remember the super user (sudo) command, or else the Raspberry Pi might think that you don't have permission to install programs.

Update temp.py

It's time to return to the *temp.py* Python program from the earlier section and beef it up with increased functionality. Fortunately, that only requires two new lines of code.

Go back to the */pi* folder you saved *temp.py* in, and open it up with the Nano text editor:

```
cd /home/pi
sudo nano temp.py
```

Just below the top section of the script, where you import the three Python libraries, you'll want to import Twilio's library:

```
from twilio.rest import TwilioRestClient
```

Below that, you'll want to insert your unique SID and Auth Token numerical strings. Put the SID where `abc` is written, and the token in place of `123`:

```
client = TwilioRestClient(account='abc', token='123')
```

Now we want to define two new variables. You will query these later in order to set conditions for when the Pi should send a text message:

```
MAX_F_TEMP = 80
MIN_F_TEMP = 70
```

Bettas are tropical fish and they like the water warm. Of course, you should update the variables depending on which range is best for your project. If you're using Celsius instead, you need to replace the F here with a C. In other words, in the following code you would use `c > MAX_C_TEMP:` and `c < MIN_C_TEMP:`.

Remember the `while` loop at the bottom of *temp.py*?

```
while True:
    print(read_temp())
    time.sleep(1)
```

Right now, it's only doing one thing: printing the temperature to the screen every second. But you want to give it some additional SMS functionality:

```
while True:
    c, f = read_temp()

    if f > MAX_F_TEMP:
        client.messages.create(to='+17035555555',
            from_='+12025555555',
            body="fish tank too hot!!")
            time.sleep(500)

    if f < MIN_F_TEMP:
        client.messages.create(to='+17035555555',
            from_='+12025555555',
            body="fish tank too cold!!")
            time.sleep(500)

    # print(read_temp())
    time.sleep(1)
```

Here, I've replaced my personal cell phone number (starting with 703) and my Raspberry Pi's cell phone number (starting with 202). I've kept the area codes so you can tell them apart from one another.

This program is telling the Pi to send a text message if the temperature exceeds the higher or lower variable specified at the top of the program. You can change the message it sends by altering what's inside the quotations on the body line. You can change the time between messages with the `time.sleep` number. Right now it is programmed to send a text every 8 minutes (500 seconds) for as long as the fish tank is experiencing a dangerously high or low temperature.

I've also commented out the `print` line because we no longer need it to write to the screen. That part was really just for testing purposes.

Paste these code snippets into *temp.py* and press Ctrl-X to save and quit Nano.

Set the maximum and minimum variables to room temperature—or artificially heat up or cool down the thermometer—and wait for your text message.

Here's what the complete *temp.py* program now looks like:

```
import os
import glob
import time

from twilio.rest import TwilioRestClient

client = TwilioRestClient(account='abc', token='123')

MAX_F_TEMP = 80
MIN_F_TEMP = 70

os.system('modprobe w1-gpio')
os.system('modprobe w1-therm')

base_dir = '/sys/bus/w1/devices/'
device_folder = glob.glob(base_dir + '28*')[0]
device_file = device_folder + '/w1_slave'

def read_temp_raw():
    f = open(device_file, 'r')
    lines = f.readlines()
    f.close()
    return lines

def read_temp():
    lines = read_temp_raw()
    while lines[0].strip()[-3:] != 'YES':
        time.sleep(0.2)
        lines = read_temp_raw()
    equals_pos = lines[1].find('t=')
    if equals_pos != -1:
        temp_string = lines[1][equals_pos+2:]
        temp_c = float(temp_string) / 1000.0
        temp_f = temp_c * 9.0 / 5.0 + 32.0
        return temp_c, temp_f

while True:
    c, f = read_temp()

    if f > MAX_F_TEMP:
        client.messages.create(to='+17035555555',
            from_='+12025555555',
            body="fish tank too hot!!")
            time.sleep(500)

    if f < MIN_F_TEMP:
        client.messages.create(to='+17035555555',
            from_='+12025555555',
            body="fish tank too cold!!")
```

```
            time.sleep(500)

        time.sleep(1)
```

Figure 4-9 shows a text that was sent from the Raspberry Pi to my phone. Ignore the first message shown here. I was experimenting with some Twilio functions before I took this.

Figure 4-9. *Cell phone receiving an alert from the Pi*

To get this text without (obviously) overheating my aquarium, I set the "maximum temp" to 50 degrees, but only for a moment.

Now the smart thermometer is monitoring the situation and alerting me at the first sign of trouble. Of course there's more you can do with all of the data pouring out of the thermometer, as you'll see next.

Building the Pi Web Server

Currently, the Python program is taking data the thermometer outputs and sharing it only in case of emergency. If you could keep tabs on the temperature fluctuations over time, even in a nonemergency situation, you might be able to predict and prevent a dramatic temperature fluctuation before it starts.

Let's build a website that displays a line graph of the temperature trend over the last 24 hours. You'll turn the Raspberry Pi—the same one you're already using to record the temperature—into its own web server to host it. To do this you will set up a MySQL database on the Raspberry Pi to store large amounts of temperature data, feeding it from the Python program to the web server.

Set Up Apache

The first step is to turn the Pi into a machine capable of hosting websites by using Apache, a free, open source HTTP web server application.

Every website you've ever visited has been hosted by one server or another, and probably that server is Apache. In fact, as of this writing, more than 50 percent of sites on the Web are hosted by servers using Apache.

Here's how it works: the World Wide Web consists of lots of URLs (Uniform Resource Locators), which specify a protocol (usually HTTP), a server (like Apache), and a URL path (*site/index.html*).

When you want to access a website, you type a URL path into your web browser. When you do that, what you're actually doing is prompting the browser to make a request to that URL's server so you can access it.

So, in order to access a website on your Raspberry Pi via a web browser, you must install a server.

Fortunately, this is a one-step process. Go to the command line and type:

```
sudo apt-get install apache2 php5 libapache2-mod-php5
```

In this command prompt, several things are happening at once. You're installing the latest version of Apache, but you're also installing two other packages: PHP and a library that helps Apache work together with PHP.

For a basic HTML site that remains static and doesn't have many features aside from text, you do not need PHP. But for an HTML site, which connects to a database and graphically displays data from it, you need a web framework. PHP is a web framework that adds more functionality to basic HTML websites.

When Apache is finished installing, restart it with this command so you can use it:

```
sudo service apache2 restart
```

Make a Basic Website

As soon as the Raspberry Pi finishes processing the `sudo service apache2 restart` command, you have a basic, working website.

Check it out. Go to your web browser and type in your Pi's local address (something like 192.168.X.X), and a very basic site should appear, headlined with the phrase, "It works!" (as shown in Figure 4-10).

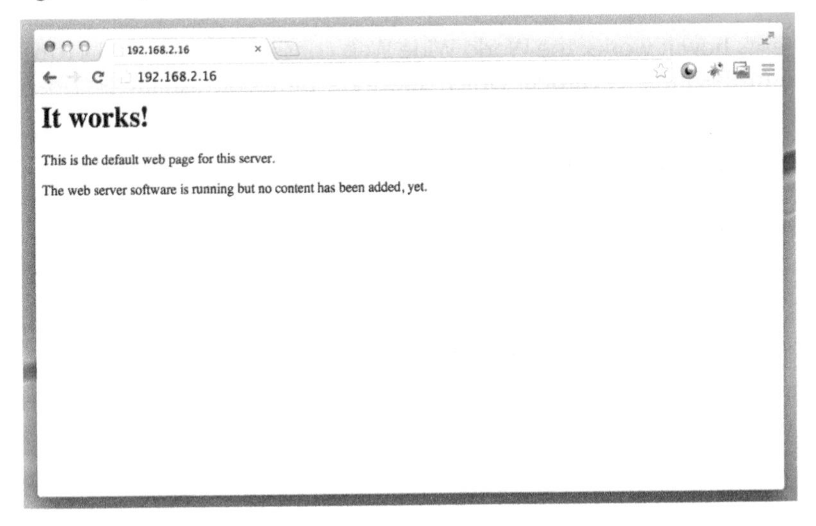

Figure 4-10. *An index.html page right after Apache is installed*

Want to tweak it? Visit the *index.html* page on your Pi:

```
cd /var/www/
sudo nano index.html
```

This simple *index.html* page came preinstalled along with Apache. Try changing the words around, saving the file, and navigating back to the Pi's local address again to watch your changes take form.

I changed the title to "Temperature Chart Over Time" because that's what this page will eventually show (Figure 4-11).

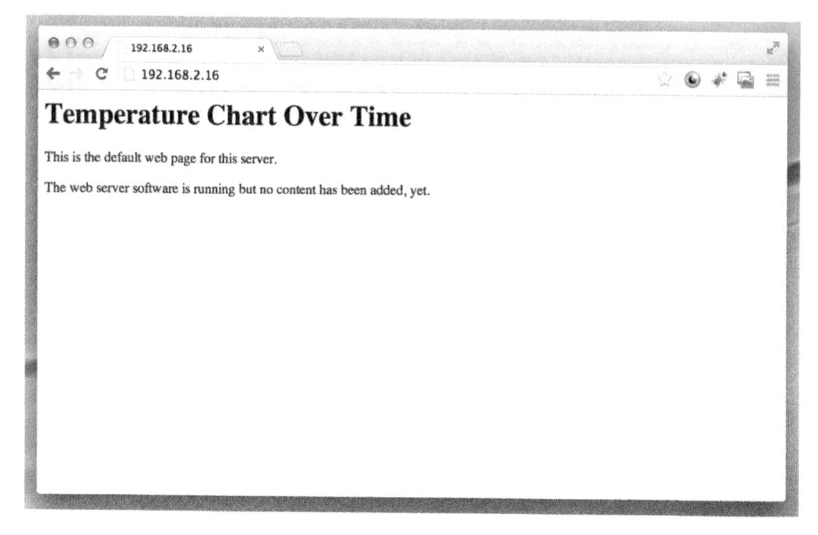

Figure 4-11. *Index.html showing changes*

Install MySQL

You now have a functioning web server, but it still isn't powerful enough to do anything with the temperature data you've been collecting. The missing link is MySQL, a free, open source database that will feed the Pi's temperature data to your web server.

Install MySQL in much the same way you installed Apache with this command:

```
sudo apt-get install mysql-server mysql-client php5-mysql
```

Once again, notice you're installing a PHP package. PHP will be the connecting framework between the MySQL database and the graphical display on *index.html*.

When you're installing MySQL, it will ask you for a username and password. Keep the username as "root," and make the password something easy to remember, since you'll need to type it in every time you want to access your MySQL database (Figure 4-12).

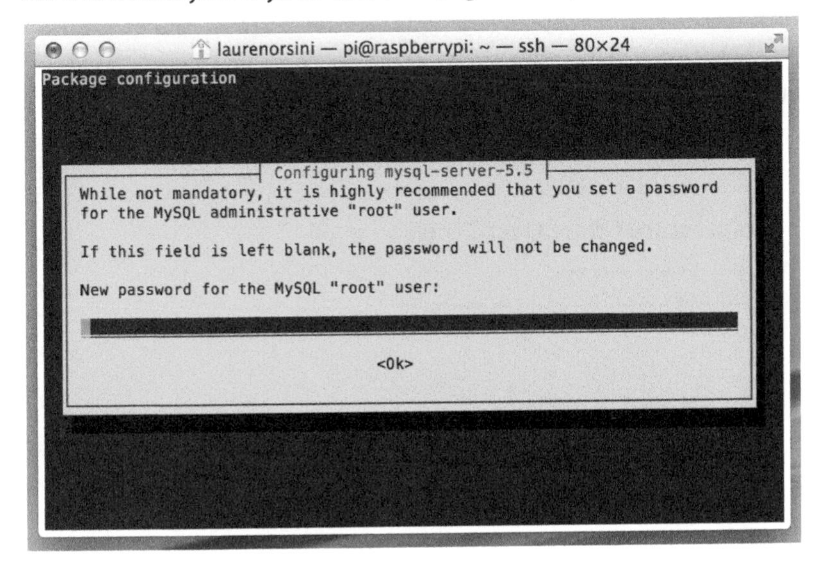

Figure 4-12. *Keep MySQL username as "root"*

You also need to install a module to let Python interact with MySQL. The information you'll be putting into the database will be coming from the *temp.py* program, and MySQL doesn't automatically interface with Python programs without a little help.

This is the command to install the Python/MySQL module:

```
sudo apt-get install python-mysqldb
```

Build a Database

Now it's time to use MySQL to build a database. Since you made the installation dependent on a username and password, this is how you get in:

```
mysql -u root -p
```

With this command, you're establishing the user as "root," and asking for a password prompt. The Pi will spit back a request for you to enter your password (Figure 4-13).

```
laurenorsini — pi@raspberrypi: ~ — ssh — 82×15
pi@raspberrypi ~ $ mysql -u root -p
Enter password:
Welcome to the MySQL monitor.  Commands end with ; or \g.
Your MySQL connection id is 43
Server version: 5.5.37-0+wheezy1 (Debian)

Copyright (c) 2000, 2014, Oracle and/or its affiliates. All rights reserved.

Oracle is a registered trademark of Oracle Corporation and/or its
affiliates. Other names may be trademarks of their respective
owners.

Type 'help;' or '\h' for help. Type '\c' to clear the current input statement.

mysql>
```

Figure 4-13. *Logging in to MySQL*

After you input your password, it's time to get building. You'll know you're inside the MySQL interface because your command-line prompts will now begin with `mysql >`.

```
CREATE DATABASE fishtank;
```

Name it whatever you want—it's just the identifier. Just don't forget that semicolon at the end that is so signature to working in MySQL!

Exit MySQL with the \q command. Enter again and type:

```
SHOW DATABASES;
```

"Fishtank" should be one of the databases that appear (Figure 4-14).

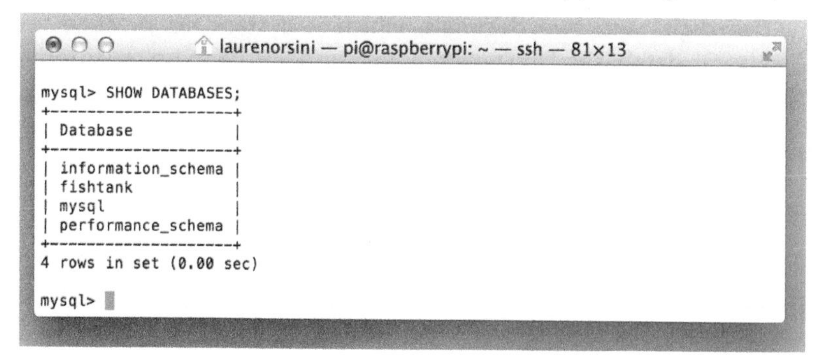

Figure 4-14. *Showing existing databases in MySQL*

Since you want to add a table to this database, type:

```
use fishtank
```

Creating a database was the easy part. It's just a storage box. The real meat of any MySQL database are the tables stored in it. So, you're going to create a table specifically to store the temperature and time values.

To build the table, type each of these prompts in order, pressing Enter in between to begin a new line. Don't forget the semicolon!

```
mysql> CREATE TABLE temperature(id INT NOT NULL PRIMARY KEY AUTO_INCREMENT,
    -> measurementTime DATETIME,
    -> temperature DECIMAL(5,2));
```

There's a lot going on here. Let's break it down:

CREATE TABLE temperature
: This builds a new table named `temperature`.

id INT NOT NULL
: This specifies that this table's values will take an integer form, and spaces for those integers cannot be left blank.

PRIMARY KEY AUTO_INCREMENT
: This assigns a unique ID to each data row in the table.

measurementTime DATETIME
: This variable, `measurementTime`, shows the date and time.

temperature DECIMAL(5,2));
: You want this variable, `temperature`, to be recorded in values with five digits and two decimals, which is a standard practice for MySQL databases. Don't expect the temperature to be so extreme that it requires more than five nondecimal digits to express!

It's a lot to take in, but what you really need to know is that you made a new table in your database named `temperature`, and added two variables for it to track: `measurementTime` and `temperature`.

Figure 4-15 shows what the previous few lines of input might look like on the command line. In other words, if you at first forget the semicolon, like I did, all is not lost.

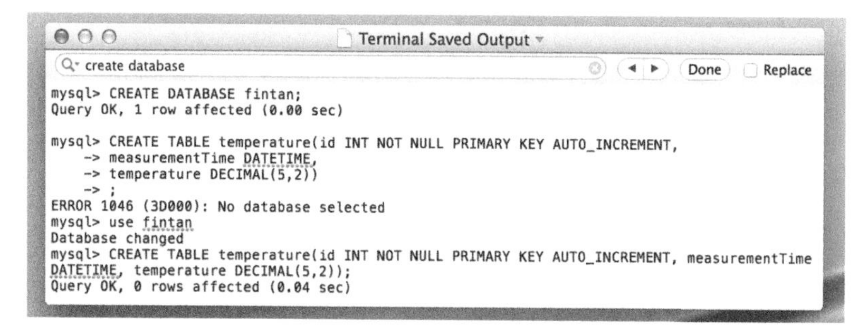

Figure 4-15. *Building the MySQL database table*

You're done here for now. Exit MySQL with the command:

```
\q
```

Connect the MySQL Database to temp.py

You now have a database to store both the temperature recording and the time at which it was recorded. But it isn't connected to your data, so it's empty right now.

Change directories to the */projects* folder and open *temp.py* with:

```
sudo nano temp.py
```

Remember how I said Python and MySQL don't automatically work together? Let's import some libraries that will allow this program to communicate with the database:

```
import MySQLdb
import datetime
```

MySQLdb is the Python library installed earlier. `datetime` is a Python library that will understand the dating format established for the `measurementTime` variable.

You need to write a function to tell *temp.py* to send the temperature data to the MySQL database table that you've built named `temperature`. Paste the following right above the `MAX_F_TEMP` variable defined in "A Thermometer That Reads and Writes" on page 132:

```
db = MySQLdb.connect("localhost", "root", "password", "fishtank")
```

```
def sendToSQL(inConn, tempVal):
        now = datetime.datetime.now()
        query = "INSERT INTO temperature (measurementTime, temperature)
VALUES ('" + now.strftime("%Y-%m-%d %H:%M:%S") + "', " + str(tempVal) + ")"
        cursor = inConn.cursor()
        cursor.execute(query)
        inConn.commit()
```

First, you're defining the db variable as the specific MySQL database. You're telling it which user, password (replace this with your actual password!), and database name to look for.

Then, you're telling the program to insert the data in the specified format. You want the program to deliver the date in the datetime format that you asked for. You want it to deliver the temperature value as it is already defined in the program—that works for your database, too:

```
while True:
    c, f = read_temp()
    sendToSQL(db, f)
    time.sleep(60)
```

Note the two new lines you're adding after the `read_temp()` function. You've created a new function named sendToSQL, which was defined earlier as a connection to the specific MySQL database. You've also told the function to do this every 60 seconds. You can adjust this depending upon how much data you want.

Finally, at the very end of the program write:

```
db.close()
```

It's important to close the database when you're no longer using it. The connection between the program and the database will otherwise keep running even after the program ends and it is no longer needed. There's no reason to use up the Pi's memory if you can help it.

Here's the whole, completed *temp.py* program one last time. You won't be editing it anymore:

```
import os
import glob
import time
import MySQLdb
import datetime

from twilio.rest import TwilioRestClient
```

```
client = TwilioRestClient(account='abc', token='123')
db = MySQLdb.connect("localhost", "root", "password", "fishtank")

def sendToSQL(inConn, tempVal):
        now = datetime.datetime.now()
        query = "INSERT INTO temperature (measurementTime, temperature)
VALUES ('"+ now.strftime("%Y-%m-%d %H:%M:%S") + "', " + str(tempVal) + ")"
        cursor = inConn.cursor()
        cursor.execute(query)
        inConn.commit()

MAX_F_TEMP = 80
MIN_F_TEMP = 70

os.system('modprobe w1-gpio')
os.system('modprobe w1-therm')

base_dir = '/sys/bus/w1/devices/'
device_folder = glob.glob(base_dir + '28-*')[0]
device_file = device_folder + '/w1_slave'

def read_temp_raw():
    f = open(device_file, 'r')
    lines = f.readlines()
    f.close()
    return lines

def read_temp():
    lines = read_temp_raw()
    while lines[0].strip()[-3:] != 'YES':
        time.sleep(0.2)
        lines = read_temp_raw()
    equals_pos = lines[1].find('t=')
    if equals_pos != -1:
        temp_string = lines[1][equals_pos+2:]
        temp_c = float(temp_string) / 1000.0
        temp_f = temp_c * 9.0 / 5.0 + 32.0
        round(temp_c,2)
        round(temp_f,2)
        return temp_c, temp_f

while True:
    c, f = read_temp()
    sendToSQL(db, f)
    time.sleep(60)
    if f > MAX_F_TEMP:
        client.messages.create(to='+17055555555',
                from_='+12025555555',
                body="fish tank too hot!!")
        time.sleep(500)
```

```
        if f < MIN_F_TEMP:
            client.messages.create(to='+17035555555',
                    from_='+12025555555',
                    body="fish tank too cold!!")
            time.sleep(500)

    db.close()
```

Once you save and close *temp.py*, it should be successfully sharing data with the temperature table.

Run the program with `sudo python temp.py`. The Pi won't actually return anything but blank space. That's because you no longer have a `print` function anywhere in this program, so it's no longer printing the results to the screen. And let's be honest, that was getting a bit obnoxious anyway.

You'll want to keep the Python program running in the background while you continue to work on other things. Otherwise, if you shut it down, it will stop collecting data.

To do this, use a program called Screen. Install it with:

```
sudo apt-get install screen
```

Once it's installed, you can run it simply by typing `screen` at any time.

Now that Screen is running, run *temp.py* as usual with `sudo python temp.py`.

temp.py is running, but while it's doing this, the window is currently occupied. So to get it to run in the background, detach it by pressing Ctrl-A-C (the "control," "a," and "c" keys on your keyboard).

It looks like a new blank window has opened and *temp.py* is gone, but it's still running! In order to check, type:

```
screen -ls
```

It'll return something like Figure 4-16.

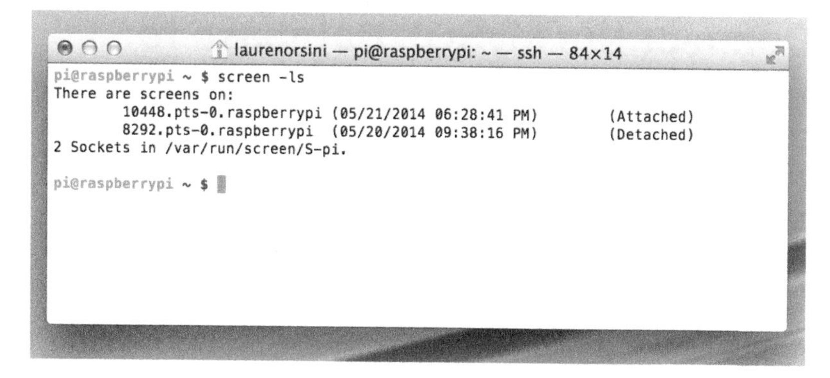

Figure 4-16. *Looking up which programs are running in Screen*

This way, you can see what's running in other windows.

Let's see how the fishtank database is responding to your changes. Log in to MySQL to check how the values are appearing in the temperature table:

```
mysql -u root -p
use fishtank
select * from temperature;
```

If all is well, something like Figure 4-17 will pop up.

Terminal Saved Output

id	measurementTime	temperature
19	2014-05-10 14:04:05	71.38
20	2014-05-10 14:04:07	71.38
21	2014-05-10 14:04:10	71.38
22	2014-05-10 14:04:13	71.38
23	2014-05-10 17:18:04	71.38
24	2014-05-10 17:18:07	71.38
25	2014-05-10 17:18:10	71.38
26	2014-05-10 17:18:12	71.38
27	2014-05-10 17:18:15	71.49
28	2014-05-10 17:18:18	72.16
29	2014-05-10 17:18:20	74.30
30	2014-05-10 17:18:23	76.10
31	2014-05-10 17:18:26	77.45
32	2014-05-10 17:18:28	78.91
33	2014-05-10 17:18:31	80.60
34	2014-05-10 17:18:34	81.84
35	2014-05-10 17:18:36	82.29
36	2014-05-10 17:18:39	82.29
37	2014-05-10 17:18:42	82.06
38	2014-05-10 17:18:45	81.84
39	2014-05-10 17:18:48	81.61
40	2014-05-10 17:18:50	81.61
41	2014-05-10 17:18:53	81.28
42	2014-05-10 17:18:56	81.16
43	2014-05-10 17:18:58	80.94
44	2014-05-10 17:19:01	80.83
45	2014-05-10 17:19:04	78.58
46	2014-05-10 17:19:07	69.35
47	2014-05-10 17:19:09	62.04
48	2014-05-10 17:19:12	57.65
49	2014-05-10 17:19:15	54.61

Figure 4-17. *A MySQL database table displaying data points*

Connect the MySQL Database to data.php

The thermometer now connects to the database. You need, however, to get the database, loaded with thermometer data, to talk to the site.

Do this by generating a new file in the same folder as *index.html*. Change your directory to the */www/* one where *index.html* is stored, and create a new blank file with Nano:

```
cd /var/www/
sudo nano data.php
```

The following is what you'll want to paste in:

```
<?php
    $username = "root";
    $password = "password";
    $host = "localhost";
    $database="fishtank";

$server = mysql_connect($host, $username, $password);
    $connection = mysql_select_db($database, $server);

    $myquery = "
SELECT  `temperature`, `measurementTime` FROM  `temperature`
ORDER BY measurementTime DESC LIMIT 10
";
    $query = mysql_query($myquery);

    if ( ! $query ) {
        echo mysql_error();
        die;
    }

    $data = array();

    for ($x = 0; $x < mysql_num_rows($query); $x++) {
        $data[] = mysql_fetch_assoc($query);
    }

    echo json_encode($data);

    mysql_close($server);
?>
```

This PHP file is connecting to the fishtank database on MySQL and telling it which pieces of data to read. In this case, you're reading the `measurementTime` and temperature variables from the temperature table.

Here's one part I want to highlight:

```
    $myquery = "
SELECT  `temperature`, `measurementTime` FROM  `temperature`
ORDER BY measurementTime DESC LIMIT 10
";
    $query = mysql_query($myquery);
```

I don't want to see every data point in the chart—that's crazy. Over time, that could be hundreds of thousands of measurement readings. So here, I've configured the table to only deliver the most recent 10 readings.

Remember how in *temp.py* I told the `sendToSQL` function to deliver a measurement reading once every hour? Combined with this PHP file, that means the graph I'm building will show the temperature trends over the past 10 minutes. So if you want to show a different time period, you can customize these two numbers.

For example, if you want to show 1 week and measure once a day, you could tell *temp.py* to send a data point once every 24 hours with `time.sleep(86400)`, and tell *data.php* to set `LIMIT 7` instead of 10.

But back to the project at hand. If you visit the Pi's local address now, you won't see a change on *index.html*. But navigate to *192.168.X.X/data.php* (replace the Xs with your unique address), and you'll see something like Figure 4-18.

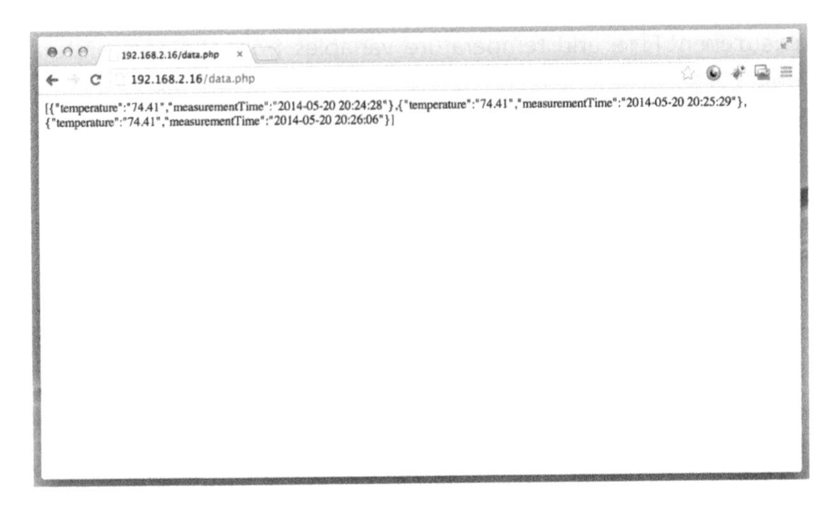

Figure 4-18. *Data points displaying on data.php*

The reason I have so many data points already is because I told *temp.py* to send a reading once per minute. So even if you want to eventually set it to once per hour or something like that, set it low initially for testing purposes.

If you see something like [ ], your MySQL database is set up correctly but it isn't collecting data, so be sure to check *temp.py* for errors. If you see the numbers, however, it's collecting the data from the MySQL database and passing it on to the site.

Design a Graph

Now the temperature data is successfully getting transferred from the thermometer, to the database, and then on to the website. But if you go to *index.html*, it's still just a blank page.

The last piece of the puzzle is to design a line graph that displays the collected data.

For this project, I used D3.js, a JavaScript library for building and displaying nice-looking data-driven documents in JavaScript, which is easy to use.

The entire script that makes *index.html* run is pretty long, so let's break down the pieces that make it work.

```
d3.json("data.php", function(error, data) {
        data.forEach(function(d) {
                d.measurementTime = parseDate(d.measurementTime);
                d.temperature = +d.temperature;
        });
```

This is the function that takes the MySQL data transferred to *data.php*, and reads it from there. You're exporting the time data with a date you parse in a specific order in the HTML (to make sure it displays correctly on the x-axis) and exporting the temperature data as is.

```
<script src="http://d3js.org/d3.v3.min.js" charset="utf-8"></script>
```

Here you are importing the D3.js library. This allows the graph to parse or convey many of the functions and definitions you're using.

Notice also that a bunch of commands in the following code refer to a variable called svg. This stands for Scalable Vector Graphics, the technology D3.js is based on. SVG is what is generating the smooth vector images based on our real-time data.

The rest of the code is defining and generating different elements of the line graph, from the x- and y-axes, to the grid lines, to the line of data itself. I've labeled each section carefully. The skeleton of this code is courtesy of Malcolm Maclean of d3noob.org. This site is a really helpful resource if you're just getting started exploring D3.js.

```
<html>

<head>
<style>

/* set the CSS */

body { font: 12px Arial;}

path {
        stroke: steelblue;
        stroke-width: 2;
        fill: none;
}

.axis path,
.axis line {
        fill: none;
        stroke: grey;
        stroke-width: 1;
        shape-rendering: crispEdges;
```

```
}

</style>
</head>

<body>

<!-- load the d3.js library -->
<script src="http://d3js.org/d3.v3.min.js" charset="utf-8"></script>

<h1>Temperature Chart Over Time</h1>

<script>
// Set the dimensions of the graph canvas

var   margin = {top: 30, right: 20, bottom: 30, left: 50},
        width = 800 - margin.left - margin.right,
        height = 270 - margin.top - margin.bottom;

// Display the date and time
var   parseDate = d3.time.format("%Y-%m-%d %H:%M:%S").parse;

// Set the X and Y axes ranges
var     x = d3.time.scale().range([0, width]);
var     y = d3.scale.linear().range([height, 0]);

// Define the X and Y axes
var     xAxis = d3.svg.axis().scale(x)
        .orient("bottom").ticks(10);

var     yAxis = d3.svg.axis().scale(y)
        .orient("left").ticks(10);

// Define the line
var     valueline = d3.svg.line()
        .x(function(d) { return x(d.measurementTime); })
        .y(function(d) { return y(d.temperature); });

// Adds the svg canvas
var svg = d3.select("body")
 .append("svg")
   .attr("width", width + margin.left + margin.right)
   .attr("height", height + margin.top + margin.bottom)
 .append("g")
   .attr("transform", "translate(" + margin.left + "," + margin.top + ")");

// Get the data
d3.json("data.php", function(error, data) {
        data.forEach(function(d) {
                d.measurementTime = parseDate(d.measurementTime);
                d.temperature = +d.temperature;
        });
```

```
        // Center the line on the graph
        x.domain(d3.extent(data, function(d) {
          return d.measurementTime; }));
        y.domain([60, d3.max(data, function(d) {
          return d.temperature; })+10]);

         // Add the valueline path
        svg.append("path")
                .attr("class", "line")
                .attr("d", valueline(data));

        // Add the X Axis
        svg.append("g")
                .attr("class", "x axis")
                .attr("transform", "translate(0," + height + ")")
                .call(xAxis);

        // Add the Y Axis
        svg.append("g")
                .attr("class", "y axis")
                .call(yAxis);

    });

    </script>

    </html>
```

Paste this code into your *index.html* file and press Ctrl-X to save it. Refresh your local Pi address in the browser window, and you should see something like Figure 4-19.

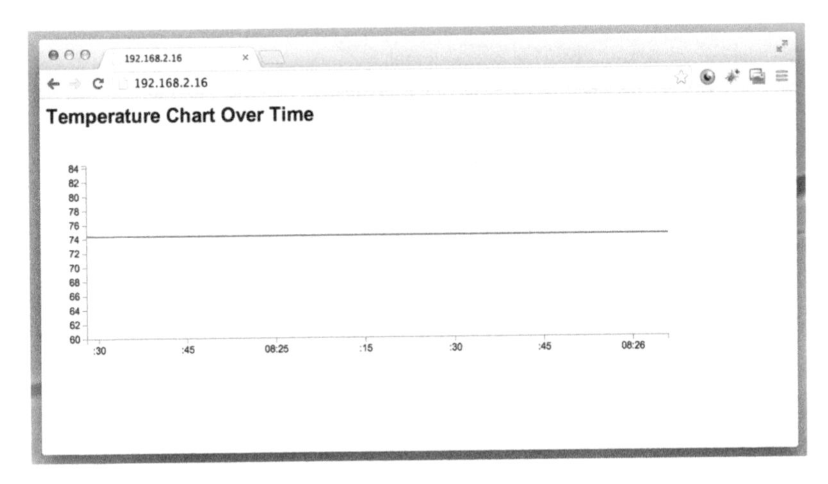

Figure 4-19. *The finished product: a graph that records the temperature over time*

This chart is kind of boring, because it's displaying only the past 10 minutes, and the thermometer has been sitting in room temperature that whole time. You can make it more interesting by lengthening the time displayed on the chart, like I explained in "Connect the MySQL Database to data.php" on page 158.

There you have it. You now have a handy chart to keep you informed of the fish tank's hourly temperature trends.

Next Steps and Acknowledgments

My fish continues swimming around his tank, oblivious to the fact that he's now living in the most high-tech aquarium setup on the block (Figure 4-20). The thermometer dipped into his water is sensing and sharing data with my Raspberry Pi, my cell phone, and my web server all at once.

Figure 4-20. *Fintan with his Raspberry Pi*

This concludes the "Internet of Fish" project, but there is still lots of room for improvement. Here are a few ways you could potentially expand on this project:

- Buy a domain name and make the local website forward facing. This is relatively easy. You can continue to host your site locally, but route the DNS for the domain name to the Pi itself. That way you can check in on the temperature even if you're thousands of miles away.

- Adjust the D3.js graph so that it shows the last week's worth of temperature readings, instead of just the last day, or the last hour's. Or use the D3.js companion program Rickshaw (*http://code.shutterstock.com/rickshaw/*) to set up an interactive toggle bar between all three.
- Set a limit on the MySQL database so it deletes data that is older than a month. Right now it's just saving *all* of the data forever, so in one year or five years, or at some point, the SD card will fill up. As it is, this program can't run indefinitely. But tweaking MySQL to delete old lines in the database would change that.

No matter what you decide to do, it's great to know that there's an entire community of Raspberry Pi tinkerers that you can rely on for advice and support. I know I wouldn't have been able to finish this project so smoothly without their help.

This section is complete but I hope it's not the last I'll hear of it. I would love for you to contact me with your finished builds, your comments and critiques, your updates and improvements, and of course, pictures of your fish.

Internet-Connected Giraffe Mood Lamp

5

By Brian Corteil

The idea for the lamp in Figure 5-1 started while walking around my local B & Q store (similar to Home Depot in the US). I spotted a child's night-light in the shape of a giraffe, and picked it up. I realized that with luck I would be able to remove the base, and use it as a diffuser for a mood lamp that I wanted to make for my wife Sarah, who loves giraffes.

I chose to use the Raspberry Pi to be the brains of this project for two reasons: the ease of connecting it to the Internet, and because it uses Python as one of its many programming languages. Apart from changing color, I wanted the lamp to be able to receive Twitter messages. I decided to design the lamp to check Twitter messages every minute or so, given that you can only poll the Twitter API with that frequency. So, the final lamp would flash when it finds a new message.

One of the things I love about this project is its versatility. You can use a handcrafted lamp shade and hide the Raspberry Pi in a thrift shop wooden box, or use a lamp shade handmade from paper, sitting on top of a laser-cut acrylic box, or use a department store table lamp. The night-light can be used to signal a number of different messages: someone is at the door, kids are home from school, a bad weather warning, temperature fluctuations, or even a sale on Make: books.

Figure 5-1. *Completed Giraffe Mood Lamp*

Materials and Tools

Following is a list of materials and tools that you need to build this lamp. You can change most of the items in this list: for example, you can use American-sized bolts and spacers instead of metric sizes. You can also go with 12V RGB strip lights instead of the LED RGB matrix, or even a box made of wood instead of the acrylic plates and spacers.

Hardware

I used acrylic plates to make the base that the lamp sits on, and cut them on a laser cutter at a local Makespace. However, you can also use 3mm plywood, or even an old wooden box from a thrift shop. If you decide to make this base, you will need the materials in Table 5-1.

Table 5-1. *Bill of materials for base*

Item	Item
3mm acrylic base plate	2x 2.5mm hex bolts, length 12mm
3mm acrylic top plate	2x 2.5mm nuts
4x PCB hex spacers 25mm	16x 3mm hex bolts, length 6mm
4x PCB hex spacers 10mm	4x sticky rubber feet
2x PCB round spacers 6mm	Double-sided sticky foam tape/pads

Electronics

The electronics are the bits that allow the Raspberry Pi to make the LEDs change color. You will need the materials listed in Table 5-2.

Table 5-2. *Electronics materials*

Item	Item
Raspberry Pi (model A or B or B+)	Push-to-make button/switch
Darlington array ULN2803A	Spotlight Kit LED RGB matrix (from UK-based Phenoptix)
Stripboard prototyping PCB, w/ 0.1" pitch hole spacing	Wire 22 AWG
Male headers strip, 0.1" spacing	2.1 jack socket, breadboard compatible
Mini WiFi adapter	10x pack of female jumpers
1x 10k resistor 1/4 watt through the hole	8GB SD card or 8GB SD card with NOOBS preloaded
1x 4-pole screw connector block for stripboard, w/ 0.1" pitch hole spacing	2mm heat shrink tube
Sparkfun Breadboard Power Supply 5V/3.3V, part number PRT-00114	9V DC 2A power supply

If you are unable to source the Spotlight LED RGB matrix Kit, you can use the materials in Table 5-3 instead.

Table 5-3. *LED matrix materials*

Item	Item
Small breadboard or small protoboard	1x 150 ohm resistor 1/4 watt through the hole
3x super bright green LEDs 20mA	1x 315 ohm resistor 1/4 watt through the hole
3x super bright red LEDs 20mA	1x 105 ohm resistor 1/4 watt through the hole
3x super bright blue LEDs 20mA	4x male to female jumpers

Selecting a Lamp

When selecting a lamp to hack, it's important to choose one made out of any material that you can safely cut. A good place to start looking for lamps is IKEA. It has a wide selection of lamps that can be hacked. Some other options could be to use a small table lamp with a hollow base to hide the Raspberry Pi in, or even to make your own lamp shade.

Tools

Table 5-4 is a basic tools list (a sample assortment is shown in Figure 5-2). The requirements can vary depending on what you are using for your lamp.

Table 5-4. *Tools list*

Item	Item
Soldering iron	Solder
Soldering iron stand	Safety glasses
Solder sucker	Screwdrivers to suit
Solder	Side cutters
Desoldering braid	Wire strippers
PCB holder, third hand, or blue-tack (poster putty)	Craft knife

Item	Item
Masking tape	Long-nose pliers
Stripboard cutter	Allen key to suit bolts

Figure 5-2. *Assortment of tools*

Table 5-5 lists optional tools for the project.

Table 5-5. *Optional tools list*

Multimeter
Hot glue gun and glue sticks
Small vise
Saw
Hand drill
Hammer
Drill bits

The multimeter is a very useful tool to have in your maker toolkit for fault finding, and checking to make sure that your circuits are working correctly. Models vary in price and can be bought from many shops, including car part suppliers, and hardware and electronic stores. In my opinion, it is worth paying more for an auto-ranging model.

Adding the Required Packages

In this section you will:

- Make sure that Raspbian is up-to-date
- Install the required Python packages (Figure 5-3)

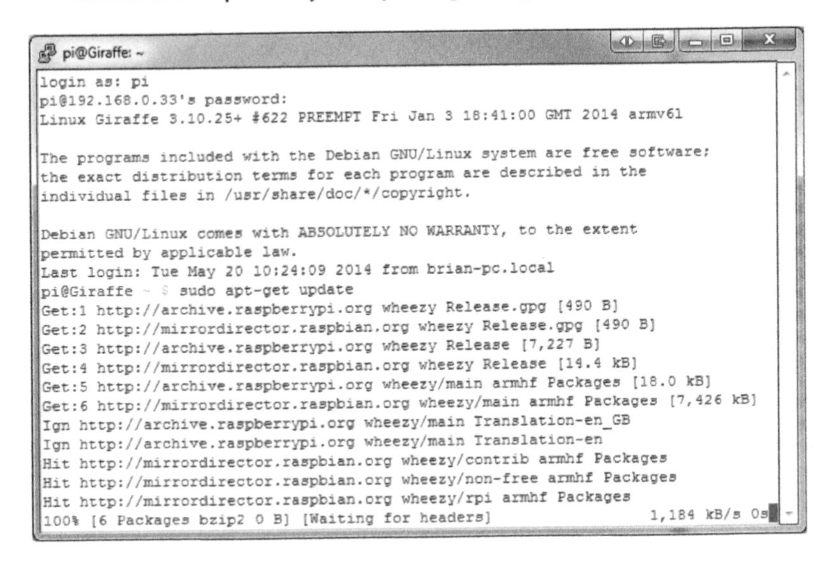

Figure 5-3. *Installing Python packages*

- Download the Pi Blaster PWM driver
- Modify the driver's code
- Install the driver's code to run when your Raspberry Pi is started

If you need help installing Raspbian or configuring your WiFi, please see Appendix A.

1. Power up, log in, and make sure that your Raspberry Pi can connect to the Internet.
2. Update your Raspberry Pi, and type the following (press Enter after each line):

```
sudo apt-get update
sudo apt-get upgrade
```

3. If there are any changes, you will be prompted. If you wish to continue, press Enter to accept, or N, and Enter to abort.

This may take up to an hour to update depending on your Internet connection and what needs to be updated. This may be a good time to start building the base.

4. Install pip, a Python package installer. This makes installing Python libraries easy and is required for the next step:

```
sudo apt-get install python-pip
```

Then press Enter when prompted.

5. Install Tweepy, a Python library that allows the Raspberry Pi to access the Twitter API and also allows the Raspberry Pi to read/send direct messages from Twitter:

```
sudo pip install tweepy
```

6. Download the *pi-blaster* source and install it. First, you need to install git core functions:

```
sudo apt-get install git-core
```

Be sure to download the source code for this project. It is located in GitHub (*https://github.com/backstopmedia/maker/*), but see Appendix A for instructions on downloading the Giraffe source code to the Raspberry Pi desktop.

7. Download the pi-blaster source:

```
git clone https://github.com/sarfata/pi-blaster.git
```

8. Change the working directory to the pi-blaster directory:

```
cd pi-blaster
```

9. Change the source code to only load three channels:

```
sudo nano pi-blaster.c
```

Change:

```
static uint8_t known_pins[] = {
    4,      // P1-7
    17,     // P1-11
    18,     // P1-12
#if REVISION == 2
    27,     // P1-13
#else
    21,     // P1-13
#endif
    22,     // P1-15
    23,     // P1-16
    24,     // P1-18
    25,     // P1-22
};
```

to:

```
 static uint8_t pin2gpio[] = {
    23,    // P1-16
    24,    // P1-18
    25,    // P1-22
};
```

10. Press Ctrl-O to save the file, press Enter, then press Ctrl-X to quit.
11. Make and install the pi-blaster daemon to run on startup:

```
sudo make install
```

In this last section, you've updated Raspbian, installed the required Python libraries, modified the pi-blaster driver source, and installed it to run when the Raspberry Pi starts.

Creating a Twitter Account for Your Lamp

In this section, you create a Twitter account and the API tokens for your lamp so you can message it.

It may be best to not complete the following steps on a Raspberry Pi. You can, but it is a little slow.

Let's get started:

1. Go to *https://dev.twitter.com/* and create an account or sign in (if you already have an account you wish to use).
2. Hover over your avatar, and click My Applications.
3. Click Create A New Application.
4. Fill out the web form. The last two fields are not that important.
5. Select the box agreeing to their terms.
6. Click Create Your Twitter Application. This page will update with your app settings. If not, hover over your avatar, click My Applications, and click the name of your app.
7. Select the Permissions tab.
8. Select Read, Write And Access Direct Messages.
9. Click Update Settings.
10. Select the API Keys tab.
11. Go to the bottom of the page and click Create My Access Token Button.
12. Open a text editor, and copy the API Key and API Secret, as well as the Access Token and Access Token Secret.

13. Save the text file, and paste/enter them into the *moodLamp.py* program at the placeholders in the Twitter account details section of the main program.

Now, return to your Raspberry Pi for the next steps. You should have created and saved the API tokens for accessing the Twitter.com service. In the next section, you will download the source code for controlling the lamp, and a program for testing the LED matrix.

Downloading Code for the Lamp and Configuring Twitter Access

In this section you download two programs: *onArg.py* is used for testing and calibrating the LED matrix, and *lamp.py* is the program for contolling the lamp. Both programs are written in Python 2.7. You will also add the API tokens and your Twitter account name to *lamp.py*.

Follow these steps:

1. Change the working directory to the user's home directory:

```
cd ~
```

2. Download *lamp.py* from git:

```
wget https://raw.githubusercontent.com/backstopmedia/maker/master/
Giraffe/lamp.py
```

3. To view the source of *lamp.py*, type the following at the command prompt:

```
nano lamp.py
```

4. You will need to change the API and access codes in the program, so that your lamp can receive Twitter messages.
5. Open the text file that contains your codes.
6. Scroll down until you find the section where the API and access tokens are stored in *lamp.py*:

```
# API keys and access tokens for twitter account
API_key = 'your API key to be placed here'
API_secret = 'your API secret to be placed here'
access_token = 'your access token to be placed here'
access_token_secret = 'your access_token to be placed here'
```

7. Copy and paste your tokens into the source code from your text file in place of the markers. If you have SSH'd into the Raspberry Pi, you can copy and paste the tokens, or type the tokens into the Raspberry Pi by hand. The tokens must be surrounded by ' or an error is created.
8. Next, find the following:

```
# twitter user name

twitterUserName = 'twitter user here'
```

9. Enter the Twitter account username in place of `twitter user here` that you wish to receive messages from. Again, the username must be surrounded by '.
10. Press Ctrl-O to save the file, press Enter, then press Ctrl-X to quit.
11. Next, test the program. The program must run as root, therefore type the following:

```
sudo python lamp.py
```

12. You should see something like Figure 5-4.

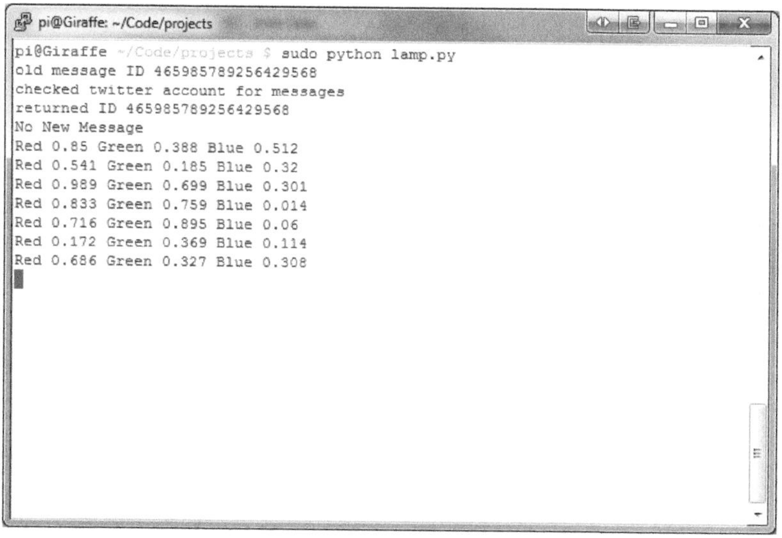

```
pi@Giraffe: ~/Code/projects
pi@Giraffe ~/Code/projects $ sudo python lamp.py
old message ID 465985789256429568
checked twitter account for messages
returned ID 465985789256429568
No New Message
Red 0.85 Green 0.388 Blue 0.512
Red 0.541 Green 0.185 Blue 0.32
Red 0.989 Green 0.699 Blue 0.301
Red 0.833 Green 0.759 Blue 0.014
Red 0.716 Green 0.895 Blue 0.06
Red 0.172 Green 0.369 Blue 0.114
Red 0.686 Green 0.327 Blue 0.308
```

Figure 5-4. *Unpacking lamp.py*

13. Send a direct message to your lamp's Twitter account. The only LED colors that have been coded in are red, green, and blue. It is possible that it may take up to 70 seconds before the lamp polls its Twitter account for new messages. You should see the new message and the text of the message on a line, as can be seen in Figure 5-5.

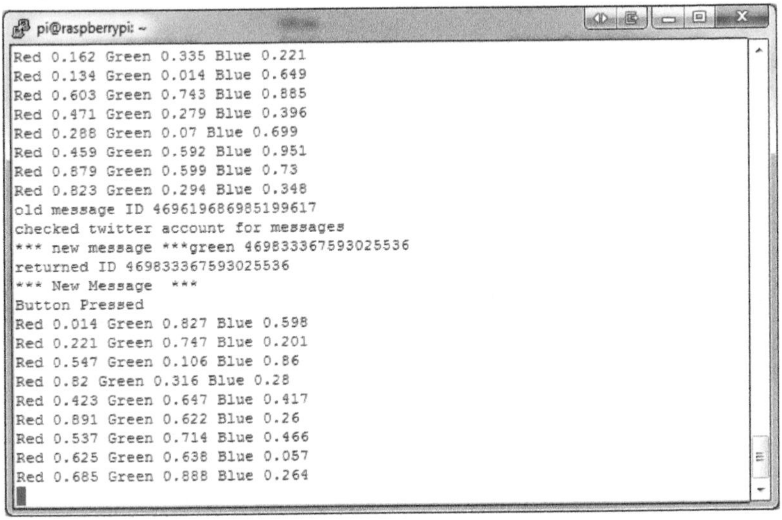

```
pi@raspberrypi: ~
Red 0.162 Green 0.335 Blue 0.221
Red 0.134 Green 0.014 Blue 0.649
Red 0.603 Green 0.743 Blue 0.885
Red 0.471 Green 0.279 Blue 0.396
Red 0.288 Green 0.07 Blue 0.699
Red 0.459 Green 0.592 Blue 0.951
Red 0.879 Green 0.599 Blue 0.73
Red 0.823 Green 0.294 Blue 0.348
old message ID 469619686985199617
checked twitter account for messages
*** new message ***green 469833367593025536
returned ID 469833367593025536
*** New Message  ***
Button Pressed
Red 0.014 Green 0.827 Blue 0.598
Red 0.221 Green 0.747 Blue 0.201
Red 0.547 Green 0.106 Blue 0.86
Red 0.82 Green 0.316 Blue 0.28
Red 0.423 Green 0.647 Blue 0.417
Red 0.891 Green 0.622 Blue 0.26
Red 0.537 Green 0.714 Blue 0.466
Red 0.625 Green 0.638 Blue 0.057
Red 0.685 Green 0.888 Blue 0.264
```

Figure 5-5. *Polling Twitter account for new messages*

You've now downloaded the source code for the lamp, added your Twitter username that the mood lamp will respond to, and added the API tokens for Twitter so that your lamp can access its Twitter account and check for new messages.

Let's next walk through the source code for *lamp.py* and *onArg.py* (the LED test program), plus learn how to use *onArg.py*.

Walking Through lamp.py

Given how crucial *lamp.py* is for our project, let's walk through what it does, and why (Figure 5-6). You will see that the first line points Linux to where the Python compiler directory is located. The next lines import the Python libraries required for the program to function.

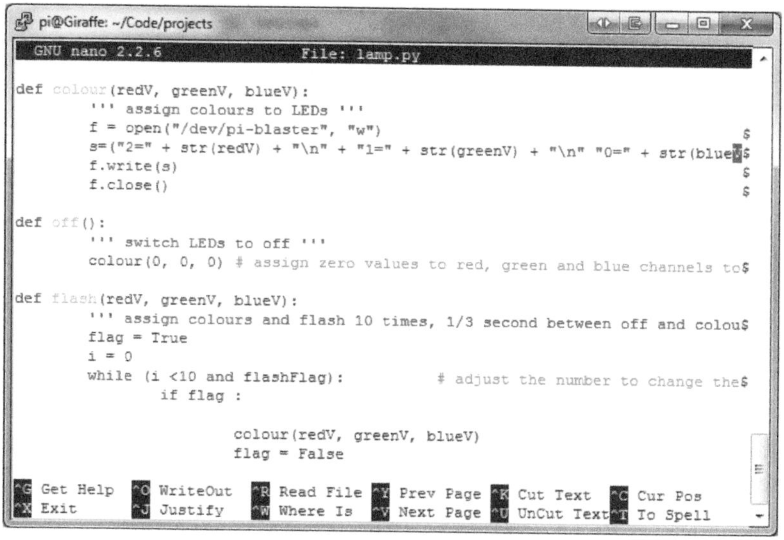

```
pi@Giraffe: ~/Code/projects
  GNU nano 2.2.6                File: lamp.py

def colour(redV, greenV, blueV):
        ''' assign colours to LEDs '''
        f = open("/dev/pi-blaster", "w")                                        $
        s=("2=" + str(redV) + "\n" + "1=" + str(greenV) + "\n" "0=" + str(blue$
        f.write(s)                                                              $
        f.close()                                                               $

def off():
        ''' switch LEDs to off '''
        colour(0, 0, 0) # assign zero values to red, green and blue channels to$

def flash(redV, greenV, blueV):
        ''' assign colours and flash 10 times, 1/3 second between off and colou$
        flag = True
        i = 0
        while (i <10 and flashFlag):          # adjust the number to change the$
                if flag :

                        colour(redV, greenV, blueV)
                        flag = False

^G Get Help   ^O WriteOut   ^R Read File  ^Y Prev Page  ^K Cut Text   ^C Cur Pos
^X Exit       ^J Justify    ^W Where Is   ^V Next Page  ^U UnCut Text ^T To Spell
```

Figure 5-6. *Lamp.py source code*

```
#!/usr/bin/python
# imports time library for sleep
import time
# imports random library for random
import random
# imports RPi GPIO library for input/output
import RPi.GPIO as GPIO
# imports tweepy library for accessing Twitter
import tweepy
# imports pickle library for saving/reading a Python object in a file
import pickle
```

Set up pin 7 on the I/O header as an input for the switch:

```
# Setup

# set pin 7 as input
GPIO.setmode(GPIO.BOARD)
GPIO.setup(7, GPIO.IN,pull_up_down=GPIO.PUD_UP)
```

Assign the API keys and access tokens to the Tweepy object, and the Twitter username:

```
# twitter account details and set up

# API keys and access tokens for twitter account
API_key = 'your API key to be placed here'
```

```
API_secret = 'your API secret to be placed here'
access_token = 'your access token to be placed here'
access_token_secret = 'your access_token to be placed here'

# OAuth process for twitter, using the keys and tokens from above
auth = tweepy.OAuthHandler(API_key, API_secret)
auth.set_access_token(access_token, access_token_secret)

# Creation of the actual interface, using authentication
twitter = tweepy.API(auth)

# twitter user name

twitterUserName = "twitter user here"
```

This section uses the pickle library to assign the saved message ID to the variable `lastMessageID` from the text file *giraffe.txt*. If you are unable to open the file, it creates a new file and sets `lastMessageID` to zero:

```
# last message ID and lastMessage string

# load lastMessageID from file giraffe.txt
# if file occurs set lastMessageID to zero
try:
    file = open('/home/pi/giraffe.txt', 'r')
    lastMessageID = pickle.load(file)
    file.close()
except IOError:
    lastMessageID = 0
    file = open('/home/pi/giraffe.txt', 'w')
    pickle.dump(lastMessageID,file)
    file.close()

lastMessage = ""
```

This sets up the flags used in the functions and main part of *lamp.py*:

```
# set flags

flag = True
messageFlag = False
flashFlag = True
```

The functions are reuseable fragments of code that are called when required. The `colour` function sets the level of the color channels red, green, and blue. It is used in all of the functions apart from itself and `my_callback`. When the `my_callback` function is called when the button is pressed, it sets the global flag `flag` to false, and prints `button pressed` to the console screen:

```
# functions
def colour(redV, greenV, blueV):
        ''' assign colours to LEDs '''
    f = open("/dev/pi-blaster", "w") # driver file to be opened
    # next line builds string to assign colours
    s=("25="+str(redV)+"\n"+"24="+str(greenV)+"\n" \
       "23="+str(blueV)+"\n")
    f.write(s)  # write s to to driver
    f.close()   # close f


def off():
    ''' switch LEDs to off '''
     # assign zero values to red, green and blue channels
     # to switch them off
    colour(0, 0, 0)

def flash(redV, greenV, blueV):
    ''' assign colours and flash 10 times, 1/3 second between off
    and colour '''
    flag = True
    i = 0

    # adjust the number after 1 to change the number of flashes
    while (i <10 and flashFlag):
        if flag :

            colour(redV, greenV, blueV)
            flag = False
        else:
            off()
            flag = True
        # adjust the number after sleep to change
        # the length of the flashes
        time.sleep(0.33)

        i = i + 1


def makespaceFlash():
    ''' example of two colour flashing using flash function '''
    off()
    flag = True
    count = 0
    # adjust the number in the while statement
    # to change the number of flashes
    while (count < 30 and flashFlag):
```

```
        if flag :                   # first colours
            red = 1
            green = 0.4
            flag = False
        else :                      # second colours
            red = 0
            green =1
            flag = True

        colour(red, green, 0)
        # adjust the number to change the length of the flashes
        time.sleep(0.33)

        count = count + 1

def messageFlash(flashColour):
    ''' selects the correct flash form the message '''
    global flashFlag        # allows access to the flashFlag flag
    flashFlag = True
    flashColour = flashColour.lower() # converts message to lower-case
    while flashFlag:
        if (flashColour == 'red'):
            flash(1,0,0)
        elif (flashColour == 'green'):
            flash(0,1,0)
        elif (flashColour == 'blue'):
            flash(0,0,1)
        else:
    # returns null for any other message and return to cycle colours
            return

def lastMessage(lastMessageID):
        ''' checks last message returns message text and
        set messageFlag to true if new message '''

    # checks messages from twitter user name
    messages = twitter.direct_messages(twitterUserName)
     # searches each message for the latest one
    for message in messages:

        messageID = message.id
        messageText = message.text
# checks to see if new message if so, prints new message to console
        if (message.id > lastMessageID):
            print ("*** new message ***" + \
                    message.text + " " + str(message.id))
            lastMessageID = message.id
            lastMessage = message
        else:
            # messages[0] is the newest message
            lastMessage = messages[0]
```

```
    # save lastMessageID to file giraffe.txt

    file = open('/home/pi/giraffe.txt', 'w')
    pickle.dump(lastMessageID,file)
    file.close()

    # returns the latest message to main program

    return lastMessage

def my_callback(channel):
    ''' button interrupt function, print button press to console when
    called and set flashFlag to False '''
        print ("Button Pressed")
        global flashFlag          # allows access to the flashFlag flag
        flashFlag = False


# end of functions
```

The next statement sets up the interrupt for the button. When the button is pressed, it calls the `my_callback` function:

```
# sets up interrupt for button
GPIO.add_event_detect(7, GPIO.FALLING, callback=my_callback,
bouncetime=300)
```

This is the main part of the program. First, it switches off the LEDs. Then, it starts a loop that never ends. The first part checks to see if there are any new messages every 75 seconds. If none are present, it prints `no new message` to the console and carries on. If there is a new message, it prints `new message` to the console, and calls the `flash` function.

The next section checks the `flag`. If true, it generates three random numbers between 0 and 1 for the three channels, and calls the `colour` function to assign them. It also assigns the current time to `start_time`. If the current time is less than 5 seconds, the `older than` start time flag is set to false, and the color change is passed by:

```
# Main program

off()
# assigns the time to message time minus 75 seconds
lastMessageTime = time.time() - 75

# main program loop

while True:
```

```
# check Twitter messages, and flash if new message

currentTime = time.time()

checkTime = lastMessageTime + 75 # must be over 70 seconds
# check if 75 seconds have passed since last check
if (checkTime <= currentTime):
    oldMessageID = lastMessageID
    print("old message ID " + str(oldMessageID))
     # set the time message was checked
    lastMessageTime = time.time()
    print ("checked twitter account for messages")
    # request latest message
    message = lastMessage(oldMessageID)
    print ("returned ID " + str(message.id))
    # assign message ID to lastMessage ID
    lastMessageID = message.id
    if (oldMessageID == lastMessageID):
        print ("No New Message")

    else:
        print ("*** New Message  ***")
        messageText = message.text
        messageFlash(messageText)


# random colours selected every 5 seconds

if (flag): # if flag is set to true update colour

    # assign random value between 0 and 1 for red,
    # green and blue channels

    red = round(random.random(), 3)
    green = round(random.random(),3)
    blue = round(random.random(),3)

    # and then set the RGB colour to the lamp

    colour(red, green, blue)

    # print out colour values to the console

    coloursValues = "Red "+ str(red) + \
                    " Green " + str(green) + \
                    " Blue " + str(blue)
    print(coloursValues)

    # start count down to colour change

    start_time = time.time()
```

```
elaspsed_time = time.time() - start_time
# change number to change time between colour changes
if elaspsed_time <= 5.0:
    flag = False

else:
    flag = True

# The End
```

onArg.py

Now, let's take a look at the *onArg.py* program. This is a test program for controlling the outputs for the lamp's RGB matrix. You can download *onArg.py* by typing the following at the command line:

```
cd ~
wget https://raw.githubusercontent.com/backstopmedia/maker/master/Giraffe/
onArg.py
```

If you run *onArg.py* without any arguments, it will turn off all of the outputs. The arguments are a number between 0 and 1. To power the red channel at 40%, green channel at 0% (off), and the blue channel at 100%, type the following at the command line:

```
python onArg.py 0.4 0 1
```

To turn off all of the channels, type the following:

```
python onArg.py
```

To use *onArg.py*, you will need to stop *lamp.py* if it is running (Figure 5-7). You do this by typing:

```
top
```

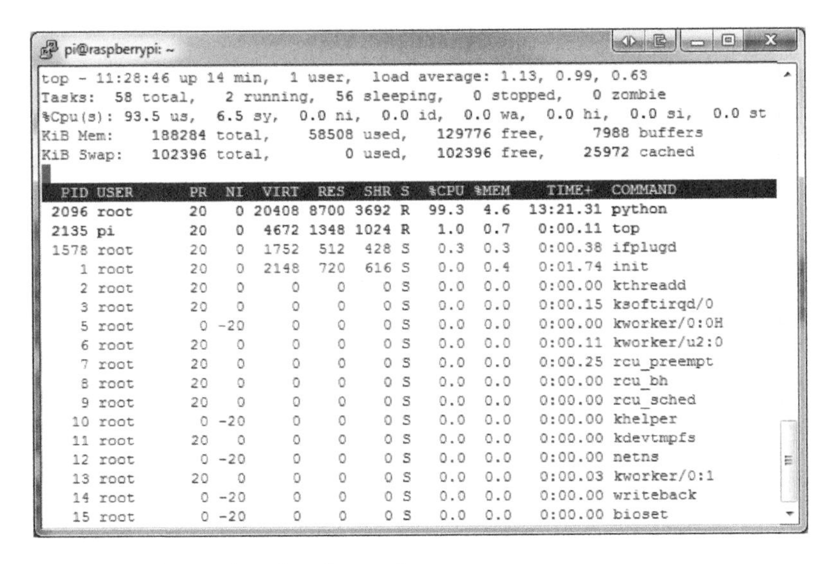

Figure 5-7. *Powering the channels*

lamp.py will be the top line under the gray bar. Make a note of its PID number. Then, press Ctrl-C to return to the command line. Enter the following command, and replace <PID> with the PID number you noted earlier:

```
sudo kill <PID>
```

Walking Through onArg.py

Given how *onArg.py* is also crucial for our project, let's walk through what it does, and why.

The first three lines are a label using the Python line comment #:

```
#
# onArg
#
```

The *sys* library is required, so the program can access the argument(s) following *onArg.py*:

```
import sys
```

The `colour` function is the same as the one from *lamp.py*:

```
# function


def colour(redV, greenV, blueV):
        ''' assign colours to LEDs '''

    # driver file to be opened
    f = open("/dev/pi-blaster", "w")
    # build string to assign colours
    s=("25=" + str(redV) + "\n" + "24=" + str(greenV) + "\n" \
       "23=" + str(blueV) + "\n")
    # write s to to driver
    f.write(s)
    # closes driver file
    f.close()
```

The main program takes the arguments and assigns them to the `colour` function, and exits:

```
# main program

# prints the number of arguments, always return one even if there
# no arguments
print 'Number of arguments:', len(sys.argv), 'arguments.'
# prints list of the arguments
print 'Argument List:', str(sys.argv)

# checks if there an argument and asigns it to the channel

if (len(sys.argv) > 1):
    redV = float(sys.argv[1])
else:
    redV = 0

if (len(sys.argv) > 2):
    greenV = float(sys.argv[2])
else:
    greenV = 0

if (len(sys.argv) > 3):
    blueV = float(sys.argv[3])
else:
    blueV = 0

colour(redV, greenV, blueV) # updates out the colour channels

# The End
```

Running lamp.py When the Raspberry Pi Starts

In this section, you edit *rc.local* (a script on the Raspberry Pi that runs when the Pi starts). This allows you to run *lamp.py* when the Pi starts.

It's best to finish the build of the base and test the LEDs before running *lamp.py*.

1. To edit *rc.local* type:

```
sudo nano /etc/rc.local
```

2. Add the following line to the end of the file:

```
python /home/pi/lamp.py
```

3. Press Ctrl-O to save the file, press Enter, and then press Ctrl-X to quit.

Now that you understand how the source code works, it's time to build the lamp.

Building the Lamp

Before building this lamp, it's important to note that some soldering of the components is required. You only need basic soldering skills to build this project.

Soldering Resource

If you've never soldered before, the following comic is a great guide. Even if you've soldered before, it's a great way to refresh your memory. You can view it at MightyOhm (*http://bit.ly/1verT7z*), or even download a copy to print out and keep.

Before you begin, it is best to have a practice soldering session. You can practice soldering spare components onto a stripboard.

From personal experience, here are some things to remember when soldering:

- Allow the soldering iron to heat up.
- Always put the iron back in its stand when you are not using it.
- Keep your iron clean, and wipe it often.
- Always solder on a nonflammable mat or surface.
- Use a PCB holder, a third hand, or a piece of blue-tack (poster putty) to hold what you are soldering.
- Heat up both the item and PCB at the same time, then add solder.
- Work in a well-ventilated area with good lighting (by an open window is ideal).
- If you are not going to use your iron for a while, switch it off.
- Use masking tape or blue-tack to hold components or wire in place.
- Start with the smaller components.

Hacking the Lamp

In this section you will remove your lamp's diffuser from its base (Figure 5-8). Let's get started.

1. Use a craft knife to carefully break the seal between the diffuser and the electronic module. See Figure 5-8.

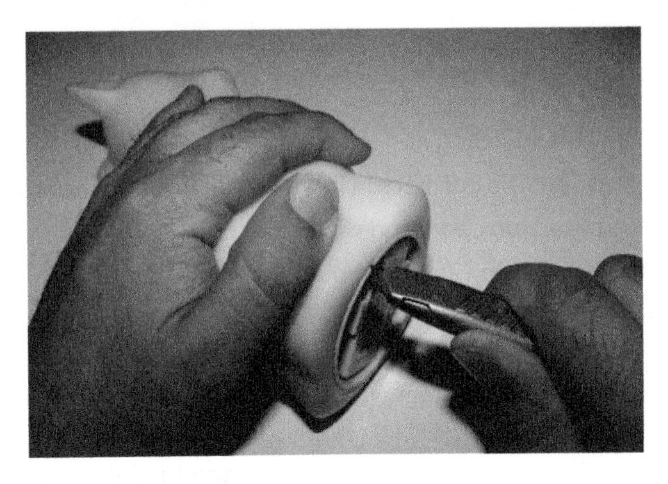

Figure 5-8. *Removing the diffuser*

2. Insert a small screwdriver in between the base (Figure 5-9).

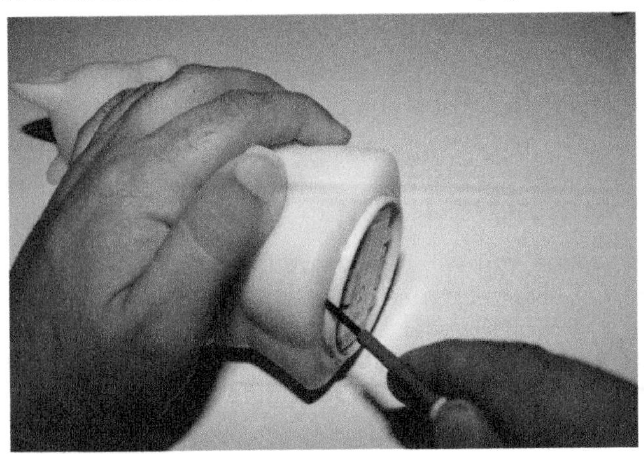

Figure 5-9. *Inserting the screw driver*

3. Use the long-nose pliers to help pull out the base (Figure 5-10).

Figure 5-10. *The diffuser pulled out*

4. Put aside the base for a future project.

DC-to-DC Converter

The mood lamp uses a 9V DC power supply to power the LEDs. If you use this to power the Raspberry Pi directly, you would destroy it. So you are going to use a DC-to-DC converter to change the 9V DC supply to the 5V DC supply that the Raspberry Pi requires. For this project, we are using a Sparkfun breadboard power supply (Figure 5-11). Let's build it.

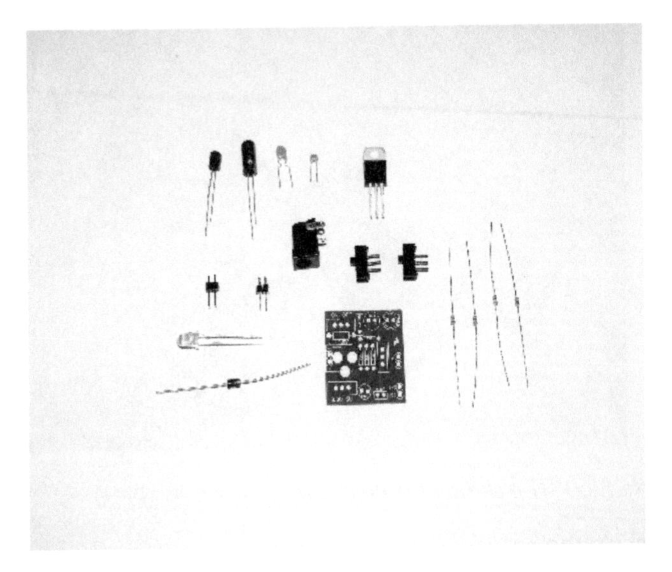

Figure 5-11. *Sparkfun breadboard power supply kit*

The diode, LED, and the electrolytic capacitors are polarized, so make sure that they are fitted the correct way.

1. The silk screen has the values of the resistors on the PCB. Place them into the correct positions. If you are not sure of the resistors' values, use a multimeter to check, or look them up with a resistor value table. Use blue-tack (poster putty) to hold them in place (Figure 5-12). Turn it over and solder. Cut off the leads when you have finished soldering them with a pair of side cutters (Figure 5-13).

Figure 5-12. *Holding the resistors in place*

Figure 5-13. *Soldered resistors*

2. Put the diode and the other small components into place. Note that the capacitors have their values printed on their sides. Use blue-tack to hold them in place, turn them over, solder, and trim the leads.

3. Solder the switches, the large capacitor, and LED into place (Figure 5-14).

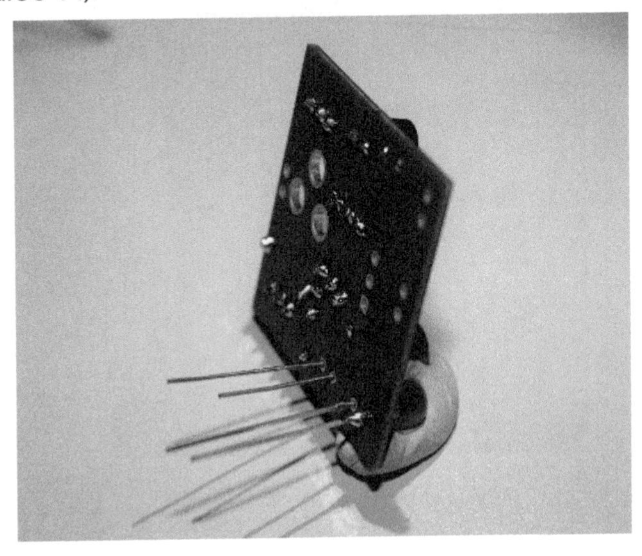

Figure 5-14. *Soldering the LEDs into place*

4. Solder the regulator into place (Figure 5-15). When the regulator is fitted correctly, the tab is facing toward the edge of the PCB.

Figure 5-15. *Soldering the regulator*

5. Solder the headers into the holes marked VCC, GND, and NU. I find it best to insert the headers into a breadboard and use some blue-tack to hold the other edge up when soldering a breakout board (Figure 5-16).

Figure 5-16. *Inserting into a breadboard*

6. Finally, solder the black wire to the – header hole, and the brown wire to the + header hole within the outline of the jack plug socket (Figure 5-17).

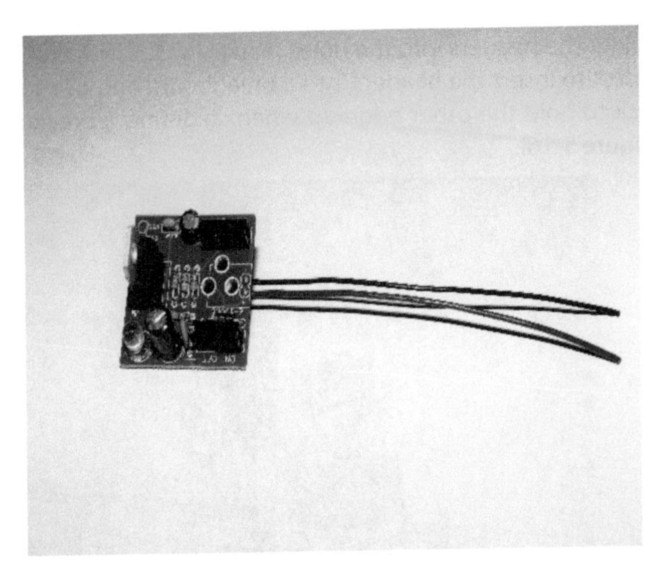

Figure 5-17. *Soldering the black and brown wires*

There is no need for the jack plug socket, so place it in your parts bin for another project.

The RGB Matrix

We are going to build our RGB matrix. If you bought an LED RGB matrix from Phenoptix (Figure 5-18), you will start with that build first. If you are building your own LED matrix on the breadboard, skip ahead to that section.

Figure 5-18. *The Phenoptix LED RGB matrix*

LED RGB matrix from Phenoptix

1. The silk screen has the values of the resistors on the PCB. Place them into the correct positions. If you are not certain of the resistors' values, use a multimeter to check, or look them up with a resistor value table. Use blue-tack (poster putty) to hold them in place. Turn it over, and solder. Cut off the leads with a pair of side cutters when you have finished soldering them.
2. Fit the LEDs into place, being careful to put the right colors in place by following the markings on the silk screen. R is for red, and so on, for the other LEDs. Once again, hold the components in place with blue-tack (poster putty), turn over, and solder them in place.
3. Next, remove 5mm off of the ends of the cable. Then, tin both ends. Solder the white cable to the + pad, and then solder the three colors to their pad. (See Figure 5-19.)

Figure 5-19. *Finished breadboard with LEDs*

Building the RGB matrix on the breadboard

Start here if you are using a solderless breadboard or protoboard.

If you are using a protoboard, solder each wire, resistor, and LED in place. Then, bridge the pads with solder to complete the circuit.

1. Place the resistors as shown in Figure 5-20.

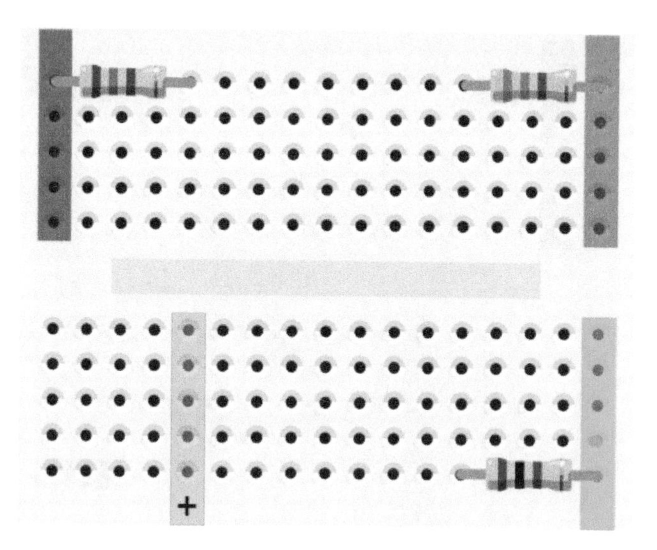

Figure 5-20. *Resistor placement*

2. Place the wire links as shown in Figure 5-21.

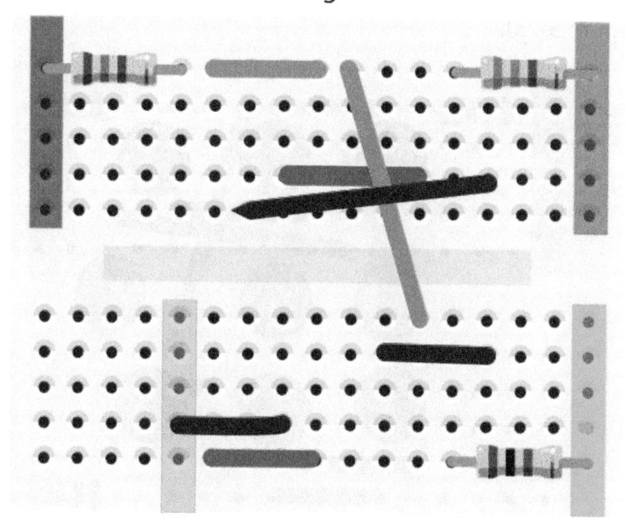

Figure 5-21. *Placement of wire links*

3. Place the LEDs onto the breadboard. The short legs go to the matching cycles with the minus symbol in them, and the long legs go to the plus symbol. Match the LED colors to the colors of the symbols (Figure 5-22).

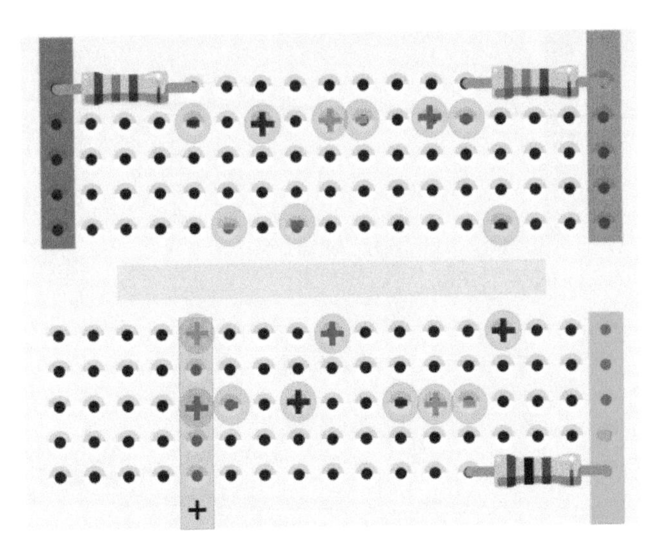

Figure 5-22. *Matching the LED legs to the + and – symbols*

4. Plug the male jump leads into each of the color columns (Figure 5-23).

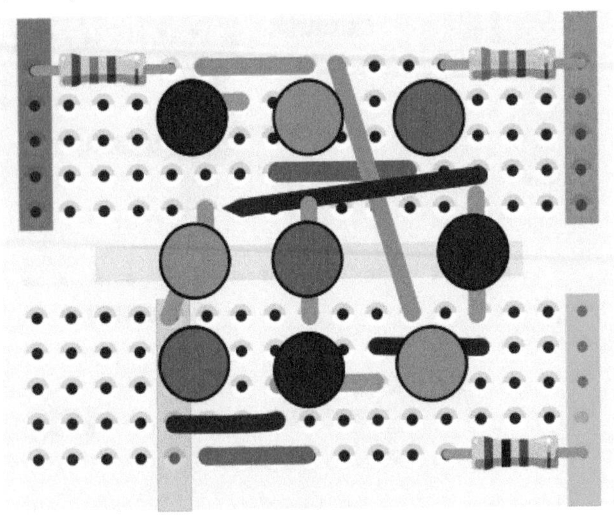

Figure 5-23. *Plug in male jump leads*

5. You can now put the finished LED matrix to one side, and start building the switch.

The Switch

The switch (Figure 5-24) is for accepting messages and connects directly to the Raspberry Pi I/O header. Let's get started.

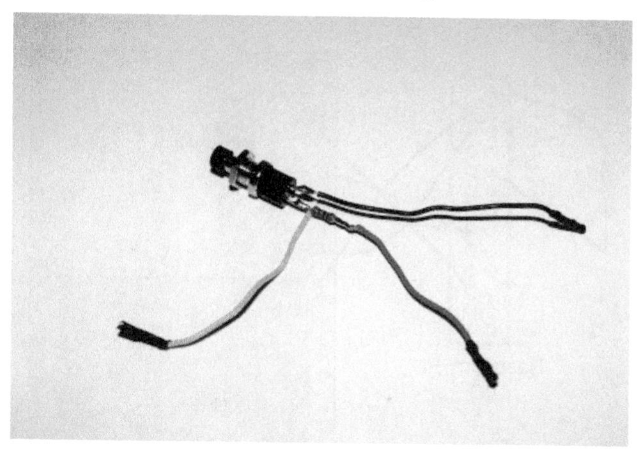

Figure 5-24. *The switch*

1. Cut three female jumpers 55mm long. It is best if they are red, yellow, and black. Strip 5mm off of each end, and tin the black jumper.
2. Tin the switch connection.
3. Solder the jumpers, as shown in Figure 5-24.

It helps to leave the leads long, and then wrap the wires around the leads. Cut them after you have finished.

4. Put the finished switch to one side, and start building the stripboard.

The Stripboard

The stripboard has two main parts: the jack plug socket with the DC-to-DC converter, and the Darlington array ULN2803 (see Figure 5-25).

The DC-to-DC converter drops the 9V DC to 5V DC for supplying the Raspberry Pi via its I/O header. The Darlington array allows the Raspberry Pi to switch the voltages and the currents up to 50V and 500mA.

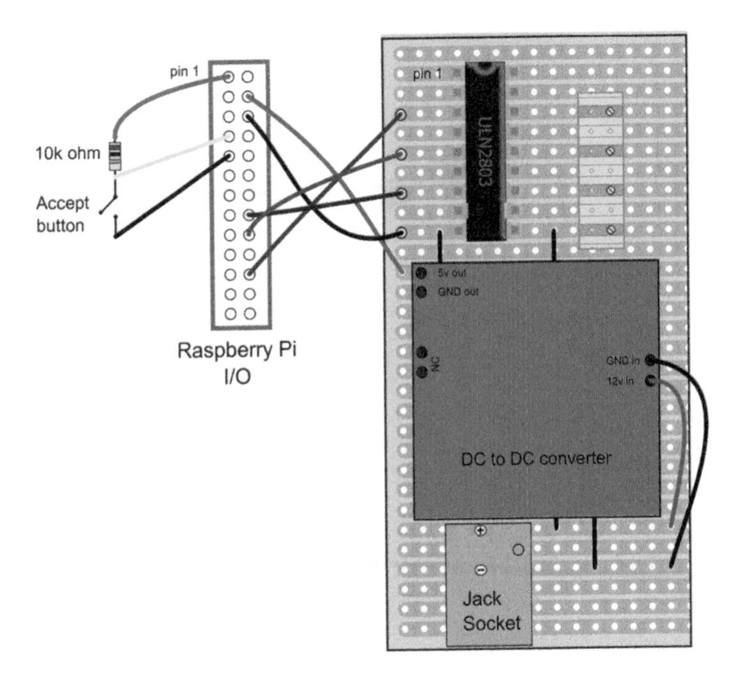

Figure 5-25. *Finished layout of the stripboard*

Always double-check that you are cutting the stripboard and setting the components into the right place.

Getting started

1. Cut the stripboard to size, 29 by 15 holes, with strips running across the shortest sides. I have found that the best way to cut the stripboard is to score it with a craft/hobby knife. Using a metal ruler as a guide so that the copper track/board is cut, break it off

in a vise by snapping it back hard. I then use sandpaper and water to sand the edges of the board.

It is easier to cut the stripboard across the holes (Figure 5-26).

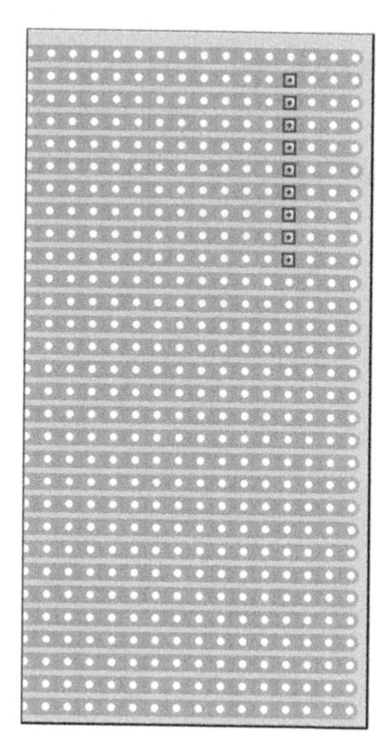

Figure 5-26. *Cutting across the holes*

2. Flip the board over so that the tracks are facing up. Cut the stripboard at the places marked. If you use a stripboard cutter, you may need to tidy up the cut with a craft/hobby knife, making sure there are no whiskers of copper joining the tracks together.

Mark the board with a marker, to show where the tracks need to be cut.

3. Cut and strip the wire for the links, removing about 5mm of insulation at each end. Bend the ends over, and place in position for each link. The links are the brown and black lines going from top to bottom of the board in Figure 5-27. Then turn over the board and solder the links into place. Trim the leads with the side cutters after soldering them.

I found that holding the wires into place with masking tape or blue-tack (poster putty) while soldering them helps.

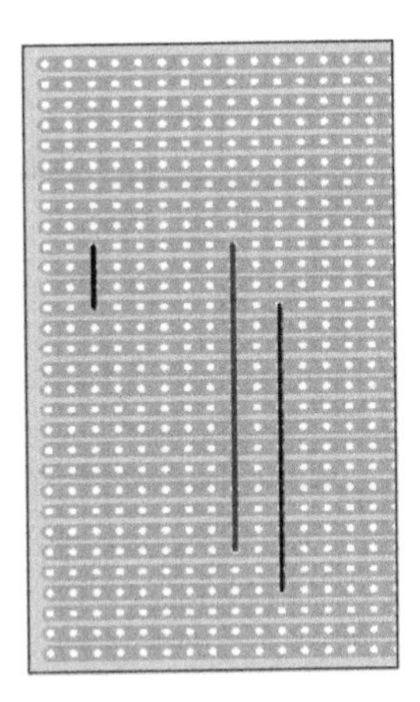

Figure 5-27. *Linking the tracks*

4. Place the Darlington array into position where you cut the tracks (Figure 5-28). Pin 1 must be in the top-left corner. A small dot or cutout marks the pin 1 end on the chip. Hold in place with blue-tack (poster putty) or tape. Then, turn over and solder in place.

It helps to mark the pin 1 position of the stripboard on the nontrack side. Solder just one pin at first, and make sure that the array is flat, and in the correct position. Reheat the joint to adjust if required. When happy, finish soldering the rest of the pins.

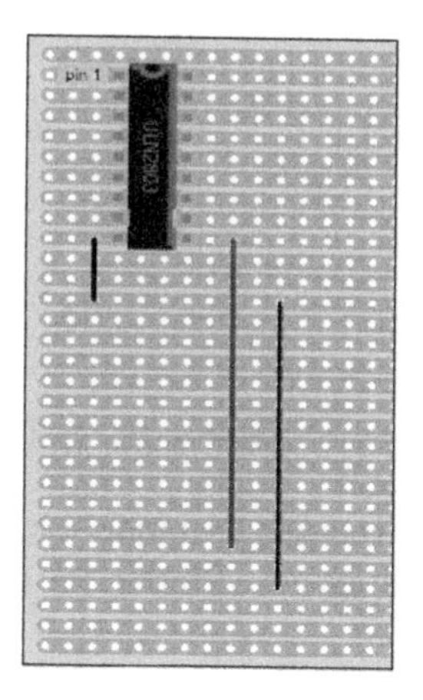

Figure 5-28. *Positioning the Darlington array*

5. Place and solder the jack plug socket in place (Figure 5-29).

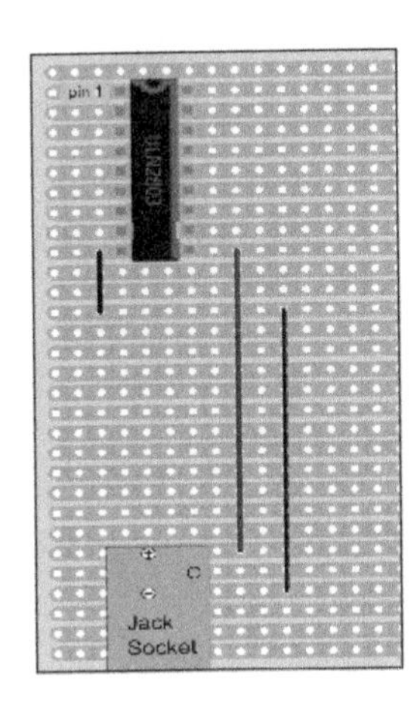

Figure 5-29. *Soldering the jack plug into place*

6. Place and solder terminal blocks into place (Figure 5-30). The bottom pin lines up with the bottom pins of the Darlington array.

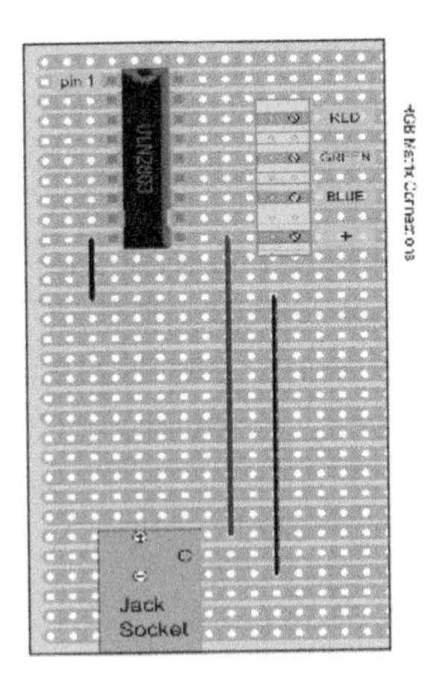

Figure 5-30. *Soldering terminal blocks into place*

7. Now, place the DC-to-DC converter in position. Strip and tin the cable ends. Place the ends into their marked positions. Be careful

that the correct polarity is observed. The two male headers should line up with both the 5V out and GND out holes on the stripboard. The other two are only for support. Solder the headers and the two wires (Figure 5-31). The brown wire goes to the 5V track, and the black wire goes to the ground track.

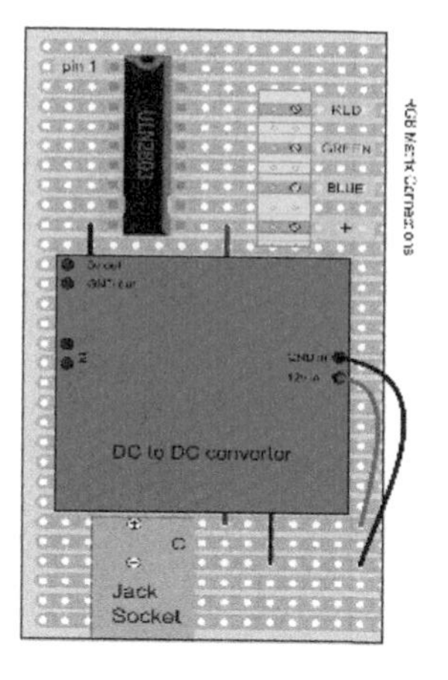

Figure 5-31. *Soldering the headers and two wires*

8. Cut five female jumper wires, about 80mm long. Strip 5mm of the cut ends, then tin and solder them into place following Figure 5-32.

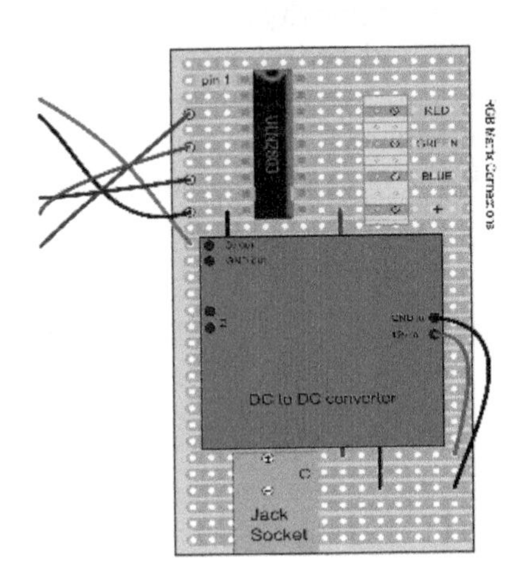

Figure 5-32. *Board with jumper wires*

9. Testing with the multimeter, select the DC voltage range if you have an auto-ranging meter, or the 20V range if not. Plug in your 12V power supply and check that about 5.2V is present at the black and brown wires. If not, check all of the connections and solder joints.

The Base

For the lamp's base, I used two acrylic plates separated by four 25mm PCB hex spacers (Figure 5-33). The bottom plate holds the Raspberry Pi and the stripboard. The top plate mounts the RGB matrix and switch. The base can be made out of almost any material, as long as it will contain the Raspberry Pi, the stripboard, and if the RGB LED matrix can sit on top of it. Again, a wooden box from a thrift shop would even work. The plans for the plates are available on this book's git repository (*https://github.com/backstopmedia/maker*).

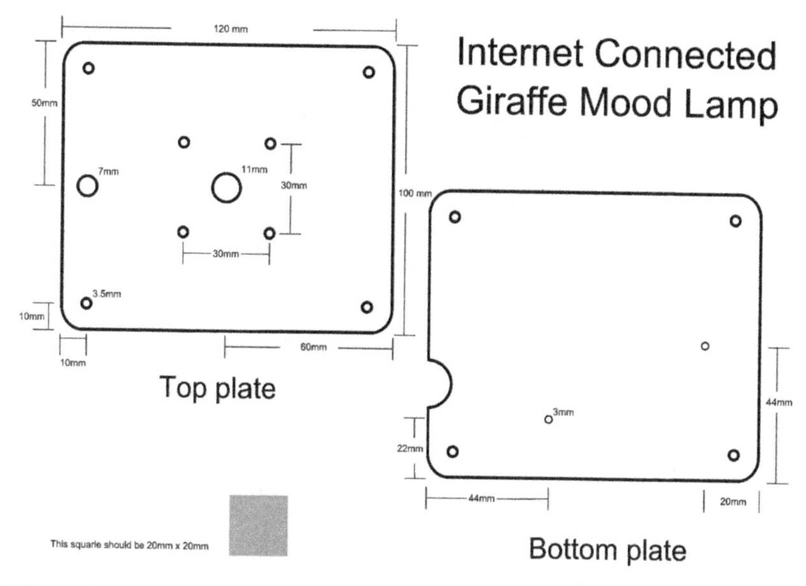

Figure 5-33. *Lamp's top plate and bottom plate*

Putting It All Together

You now have everything in place to finish building the mood lamp. So let's put it all together with the following steps:

1. Start by attaching the Raspberry Pi to the base using the 3mm spacers and the 12mm 2.5 bolts (Figure 5-34).

Take care to not overtighten the nuts. You could damage the Raspberry Pi's PCB.

Figure 5-34. *Attaching Raspberry Pi to the lamp base*

2. When the Raspberry Pi is in place, stick down the stripboard PCB with the double-sided foam tape or sticky pads, making sure that the power socket is on the same side as the USB ports (Figure 5-35).

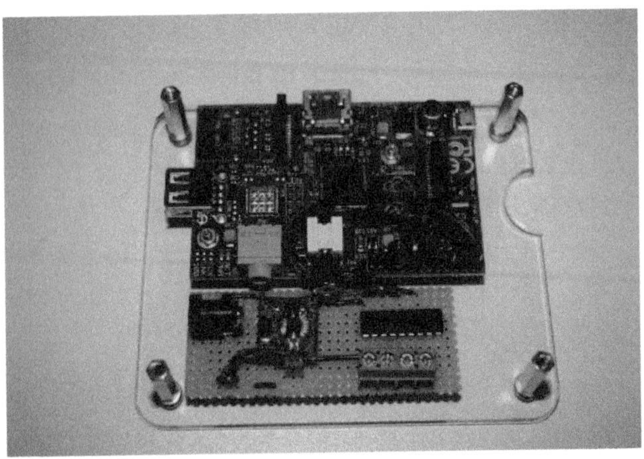

Figure 5-35. *Attaching with double-sided tape*

3. Connect the headers to the Raspberry Pi as shown in Figure 5-36.

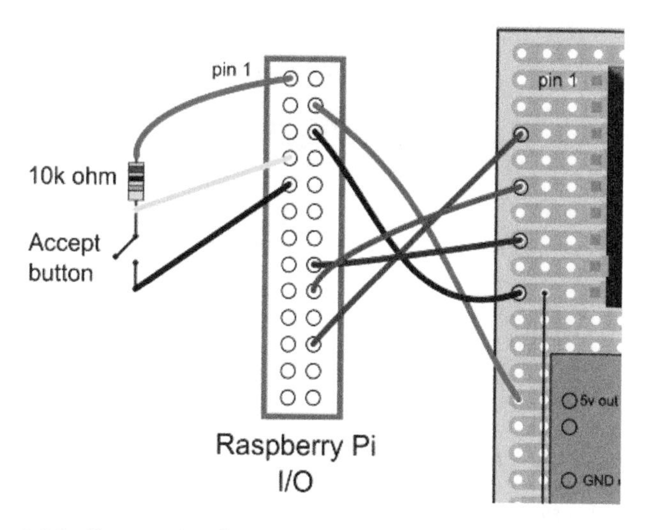

Figure 5-36. *Connecting headers to the Pi*

Make sure each plug is in the correct place. Follow the diagram carefully and check twice. An error could destroy or damage your Raspberry Pi.

If you used the same colored wires as suggested earlier, they will be:

Cable Color	Pin Number
Red	4
Black	6
Brown	25
Green	24
Blue	23

4. Using four 10mm spacers, fit the LED matrix onto the top plate using eight of the 6mm bolts (see Figure 5-37). Remember to feed the wires through the hole and connect them to the screw terminals, as shown in Figure 5-37.

Figure 5-37. *Fitting the LED matrix to the top plate*

5. Fit the switch into the hole in the top plate, and connect it to the Raspberry Pi's I/O pins:

Wire Color	Pin Number
Red	1
Yellow	7
Black	8

6. Power up the mood lamp. Wait until the LEDs are lit up and then send a message to your lamp. The lamp should flash the color you sent within 75 seconds or less, and return to cycling when the button is pressed. If the LED matrix does not light up, or the wrong colors are displayed, check that the LED matrix plugs are connected to the right pins. If pressing the button does not cancel the signal, check connections and solder the joints of the switch.

   Example 5-1.

   It's best to switch off your Raspberry Pi to avoid damaging it while you are checking to see if the LED matrix plugs are connected to the right pins.

7. Fit the top plate to the base plate, and then turn over the base and fit the feet onto the bottom of the base (Figure 5-38).

Figure 5-38. *Fitting the top and bottom plates*

8. Place the diffuser on top.
9. Grab a drink, pat yourself on the back, sit down, and enjoy watching your mood lamp!

Where to Go from Here?

Due to the power of the Raspberry Pi, this project can be expanded in a number of ways. The lamp could be made to flash when your favorite football team scores a goal or touchdown. The lamp could be made to alert you when someone is approaching your house. It could be made to react to a Twitter hashtag, change color according to the temperature inside and out, warn you when there is possible ice on the road, or even provide you with a tornado warning.

The Raspberry Pi can also host a website that allows you to change the color of the lamp, or change its settings. In addition, you could also improve the *lamp.py* program so that the change between colors is smoother. The possibilities are endless. Please let me know what you have done with your lamp. Tweet a message, picture, or video with the hashtag #GiraffeMoodLamp. I'm really looking forward to seeing what you have made!

Appendix: The Missing Pieces

Follow along to learn how to do the following:

1. Download the book's source code
2. Install Raspbian on your Pi
3. Enable WiFi on your Pi
4. Convert from a breadboard to a stripboard
5. Reassign conductive keyboard keys
6. Manually install the software for the conductive keyboard

Downloading the Book's Source Code

To download the book's source code, first install git. Open the LXTerminal window and type:

```
sudo apt-get install git-core
```

You might be asked for your password, but if your Pi has access to the Internet, git should be installed.

Now type the following in the LXTerminal window:

```
cd Desktop
git clone https://github.com/backstopmedia/maker.git
```

The source code will be installed in a Maker folder on the Pi's desktop. You will get access to the source code of the projects used in this book (Picussion, Fishtank, and Giraffe) through the following path:

/home/pi/Desktop/maker

or

~/Desktop/maker

Installing Raspbian

To install Raspbian on your Pi, you will use NOOBS, which is available from the downloads section of the Raspberry Pi Foundation's website at *http://www.raspberrypi.org/downloads*.

Select the NOOBS download. If you require help with the install, follow the setup guide in the Raspberry Pi help section at *http://www.raspberrypi.org/help/*.

Make sure that you have backed up any files on the Pi's SD card that you wish to keep before formatting it and installing NOOBS.

You can skip steps 1 through 4 if you purchased an SD card with NOOBS already installed on it.

1. Download NOOBS from the Raspberry Pi Foundation website (*http://www.raspberrypi.org/downloads/*) and select the NOOBS ZIP download or the Torrent version if you have a Torrent client.

    The NOOBS download is very large (over 1GB). If you are on a slow connection, this could take a long time to download.

2. Format your SD card before copying the NOOBS files onto it.

- Visit the SD Association's website (*https://www.sdcard.org/downloads/formatter_4/*) and download SD Formatter 4.0 for either Windows or Mac.
- Install the software.
- Insert the SD card onto your computer's or laptop's card reader and note the drive letter assigned to it (e.g., F:).
- In SD Formatter, select the drive letter for your SD card and format it.

3. When the download has finished, extract the files from the ZIP.
4. Copy the contents of the folder to your SD card. When it has finished copying the files over, safely remove the card from your card reader and insert it into the Raspberry Pi's SD holder.
5. Connect the USB hub to your Raspberry Pi.
6. Plug the mouse, keyboard, and WiFi adapter into the hub.
7. Connect your Raspberry Pi to your monitor or television.
8. Switch your monitor/television to the input that the Raspberry Pi is connected to.
9. Plug in your USB power lead, and you should see the NOOBS selection screen within a few seconds.
10. Select Raspbian. This can take about 20 minutes.
11. When the install is finished, press the Enter key.
12. Your Raspberry Pi will reboot. When it has booted, the Raspberry Pi Software Configuration tool is displayed.
13. Using the cursor keys, select Finish, and press Enter to go to the command line.

Next, you will connect to your network with the WiFi adapter.

Configuring the WiFi Adapter

In this section you will configure your WiFi adapter to connect to your home network. If you plan to use an Ethernet lead to connect to your network, you can skip this section.

1. Enter this at the command line in Raspbian on your Pi:

```
startx
```

2. This will start the GUI desktop.
3. Double-click WIFI config. The wpa_gui window will open.
4. With the mouse, select the Manage Networks tab, and click the Scan button. The Scan Results window will open.
5. Click the Scan button and double-click your home network from the list.
6. A new window will open. In the field labeled PSK, enter your network password if required, and click the Add button to save your network.
7. Close the window, and close the wpa_gui window as well.
8. Double-click the LXTermail icon.
9. When the window loads, type the following:

```
sudo shutdown -r now
```

10. The Raspberry Pi will now reboot.
11. Log on. The default username is `pi` and the password is `raspberry`.
12. Next, enter:

```
ping www.bbc.co.uk
```

13. If `ping: unknown host www.bbc.co.uk` is displayed, you have no network connection. Start from step 1 again. If the screen is scrolling, press and hold the Ctrl key and press C on Windows (on a Mac, also hold the Ctrl key and press C).
14. Your screen should look something like Figure A-1.

You have now configured your WiFi adapter to connect to your home network, and then to the Internet, and tested it all. At this point you could work with your Raspberry Pi headless. Working "headless" means using the Raspberry Pi without plugging in a keyboard, mouse, or monitor. Instead, you'll access the Raspberry Pi via either its wired Ethernet port, a wireless WiFi adapter, or through the console via a USB TTL (Transistor-Transistor Logic) Serial Debug / Console lead.

```
pi@raspberrypi ~ $
pi@raspberrypi ~ $
pi@raspberrypi ~ $
pi@raspberrypi ~ $ ping www.bbc.com
PING www-bbc-com.bbc.net.uk (212.58.246.94) 56(84) bytes of data.
64 bytes from bbc-vip015.cwwtf.bbc.co.uk (212.58.246.94): icmp_req=1 ttl=53 time
=17.7 ms
64 bytes from bbc-vip015.cwwtf.bbc.co.uk (212.58.246.94): icmp_req=2 ttl=53 time
=17.6 ms
64 bytes from bbc-vip015.cwwtf.bbc.co.uk (212.58.246.94): icmp_req=3 ttl=53 time
=18.5 ms
64 bytes from bbc-vip015.cwwtf.bbc.co.uk (212.58.246.94): icmp_req=4 ttl=53 time
=18.4 ms
64 bytes from bbc-vip015.cwwtf.bbc.co.uk (212.58.246.94): icmp_req=5 ttl=53 time
=20.0 ms
^Z
[1]+  Stopped                 ping www.bbc.com
pi@raspberrypi ~ $
```

Figure A-1. *Configuring the WiFi adapter*

You can use the open source SSH client PuTTY software, available for Windows and Linux from PuTTy (*http://bit.ly/XSB88p*) and use netscan network scanning software to find your Raspberry Pi. If you are using a Mac, the SSH functionality is built right into the OS X Terminal (no need for additional libraries or programs).

Your PC and Raspberry Pi must be on the same network and have the same subnet. If they don't, you will not be able to connect to the Raspberry Pi from your PC. Also, with a Mac make sure the Pi is on the same network as the Mac.

From Breadboard to Stripboard

The following section applies to the conductive keyboard (the ShrimpKey) from Chapter 3.

If you know how to solder (or if you want to try to learn it), you can transfer your conductive keyboard breadboard to a stripboard. This way you'll get a more permanent keyboard.

In this section, you're making a single layered stripboard version of the conductive keyboard. Further instructions can be found on Sjoerd Dirk's blog fromScratchEd (*http://fse.link/r16*).

You need two new materials:

- 1x stripboard (94 x 55 mm)
- 1x 28-pin DIP IC socket

The stripboard should have at least 26 columns and 17 rows with horizontal copper strips.

You'll also need a sharp knife and a metal ruler. First cut the strips (at the copper side of the stripboard) as shown in Figure A-2. Be careful not to cut the strips at the marked places.

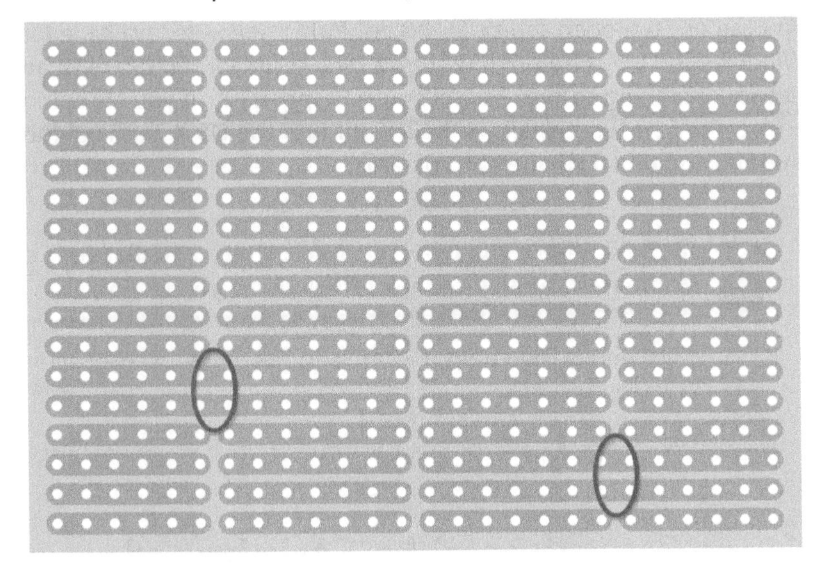

Figure A-2. *Stripboard copper side*

Now turn the stripboard over. The noncopper side should be on top.

In order to make soldering easier, the layout for the stripboard differs slightly from the breadboard layout built in Chapter 3 (see Figure A-3).

- One 68 Ω resistor is moved and is now placed exactly like the blue wire on the breadboard.
- The blue wire is removed.
- The 330 Ω resistor is moved and is now placed exactly like the purple wire on the breadboard.
- The purple wire (for the LED) is removed.
- One red wire is removed (at the right side of the chip).

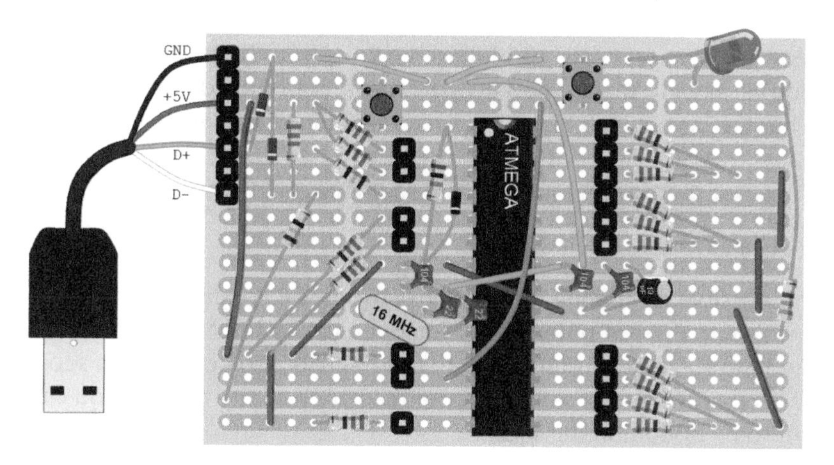

Figure A-3. *Stripboard noncopper side*

All of the components are placed on the noncopper side and soldered on the copper side.

Now you can start soldering. Start with the 28-pin DIP IC socket. Using this socket is safer. You don't have to solder the chip directly on the stripboard, so that way you can't overheat the chip.

Then place the chip in the socket and solder the other components in the following order. You would typically place the chip in the socket after the soldering is completed, however given that we are soldering across the ATmega chip, it is easier to place the chip first.

1. Diodes
2. Resistors

3. Capacitators
4. Buttons
5. Pin headers
6. Crystal
7. LED
8. Wires
9. USB cable

Now you can test your conductive keyboard on the stripboard. Since you've already made the keyboard on a breadboard (refer back to Chapter 3), you don't have to upload the ShrimpKey sketch again.

Reassigning Conductive Keyboard Keys

It's possible to change which signals from the conductive keyboard are sent to the computer. When you open the ShrimpKey sketch in the Arduino IDE, a second file (called *settings.h*) is also opened in a separate tab. To change the behavior of the conductive keyboard, you can set some options in this file.

The numbering of the inputs is as shown in Figure A-4 (refer to Chapter 3).

Pin number 6 (see Chapter 3) is a special pin: it's used as an output.

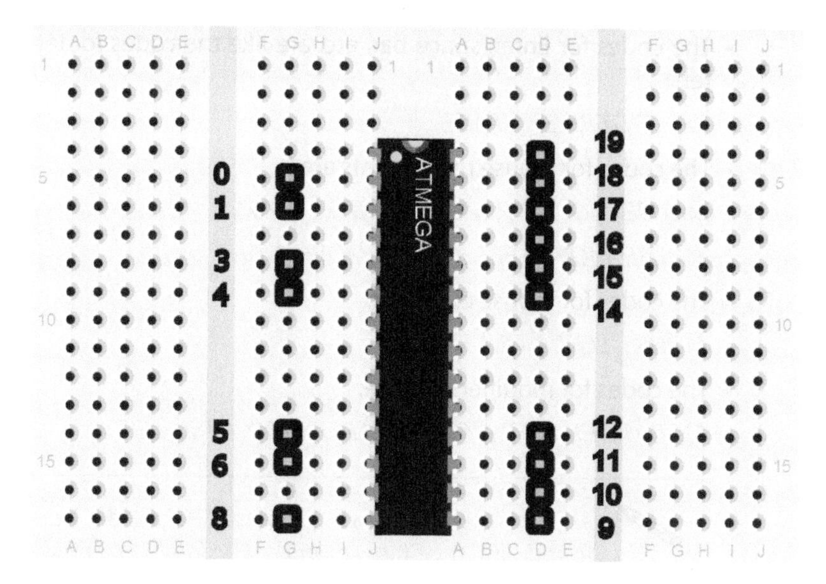

Figure A-4. *Pin numbering*

The options you can safely change are:

define NUM_MODS
: This defines how many of the inputs should be used for the modifier keys (Alt, Shift, Control).

int modPinNumbers[NUM_MODS]
: This defines which inputs are used as a modifier, separated by commas (see Figure A-4 for the numbers). The amount of defined pins in this option should match *define NUM_MODS*.

char keyCodes[NUM_INPUTS]
: This is the actual mapping between the inputs and the send signals, and always contains 16 codes. Each code is separated with a comma. The mapping is in the order of the pins, so the first code is for pin 0, the second for pin 1, etc. Make sure that you use modifier codes at the pins you defined at *int modPinNumbers[NUM_MODS]*. (All codes should be typed in capitals.)

  - The codes for letters start with KEY_, followed with the desired letter:

    The A should be defined as *KEY_A*.

- The codes for Enter, space bar, etc. are like the codes for letters:

  KEY_ENTER, KEY_SPACE
- The codes for mouse movements are:

  MOUSE_MOVE_UP, MOUSE_MOVE_DOWN,

  MOUSE_MOVE_LEFT, MOUSE_MOVE_RIGHT
- The codes for mouse clicks are:

  MOUSE_LEFT, MOUSE_RIGHT
- The codes for modifier keys are:

  MOD_SHIFT_LEFT, MOD_ALT_LEFT, MOD_CTRL_LEFT

A complete list of all supported keys can be found in the file */home/pi/arduino-1.0.5fSE/libraries/UsbKeyboard/UsbKeyboard.h.*

After changing the key definitions in the IDE, you can upload the sketch. Press the programming button on the conductive keyboard and connect the USB cable. Release the programming button once the cable is connected. Now press the upload button in the Arduino IDE and the new sketch will be uploaded.

Manual Installation

This section describes how to install the necessary software on the Raspberry Pi that is required for the Raspberries from Scratch project, and how to change some of its settings. If you already followed "Automatic installation" on page 72, you don't need to install anything here. This manual installation follows the same order as the automatic installation.

First boot up your Raspberry Pi with Raspbian.

The following commands can be typed at the command prompt or in an LXTerminal session on Raspbian. The commands assume you're logged in as the "pi" user. If that's not the case, change the commands when necessary.

Enabling SPI (Serial Peripheral Interface)

The SPI pins are used to burn the bootloader on the ATmega chip (this is explained in "Raspberry Pi as AVR Programmer" on page 72).

1. Start raspi-config:

```
cd /home/pi
sudo raspi-config
```

2. Go to "8 Advanced Options."
3. Go to "A5 SPI" and enable SPI.
4. Exit raspi-config.

Install AVRDUDE

AVRDUDE is the software used to burn the bootloader, which is explained in "Raspberry Pi as AVR Programmer" on page 72.

Type the following commands at the command prompt or in LXTerminal. These commands will return several screens full of information as the commands run. You can enter the next command when the prompt returns.

```
sudo apt-get update
sudo apt-get install bison autoconf make gcc flex
    libusb-1.0-0 libusb-1.0-0-dev -y
cd /home/pi
git clone https://github.com/sdmeijer/avrdude --depth 1
cd avrdude/avrdude
./bootstrap
./configure
make
sudo make install
```

Install Arduino IDE

The default Arduino IDE version on the Raspberry Pi is version 1.0.1, but we need version 1.0.5. Also, the AVR Toolchain (a collection of tools/libraries used to create applications for AVR microcontrollers) on the Raspberry Pi is newer than the version that Arduino included in its

IDE. In order to compile the ShrimpKey software with this newer AVR Toolchain I had to patch the default version with a small bugfix.

First, install the default Arduino IDE:

```
cd /home/pi
sudo apt-get install arduino -y
```

Then manually update this version with these commands (there are a lot of commands because it's a manual update and you have to make links using `ln` between some files):

```
wget http://fse.link/r09 -O arduino-1.0.5fSE-linux.tgz
tar xvzf arduino-1.0.5fSE-linux.tgz
cd arduino-1.0.5fSE
cd lib
ln -sf /usr/lib/jni/librxtxSerial.so librxtxSerial.so
ln -sf /usr/share/java/RXTXcomm.jar RXTXcomm.jar
cd /home/pi/arduino-1.0.5fSE/hardware/tools
ln -sf /usr/bin/avrdude avrdude
ln -sf /etc/avrdude.conf avrdude.conf
cd /home/pi/arduino-1.0.5fSE/hardware/tools/avr/bin
ln -sf /usr/lib/avr/bin/ar avr-ar
ln -sf /usr/lib/avr/bin/as avr-as
ln -sf /usr/lib/avr/bin/ld avr-ld
ln -sf /usr/lib/avr/bin/nm avr-nm
ln -sf /usr/lib/avr/bin/objcopy avr-objcopy
ln -sf /usr/lib/avr/bin/objdump avr-objdump
ln -sf /usr/lib/avr/bin/ranlib avr-ranlib
ln -sf /usr/lib/avr/bin/strip avr-strip
ln -sf /usr/bin/avr-cpp avr-cpp
ln -sf /usr/bin/avr-g++ avr-g++
ln -sf /usr/bin/avr-gcc avr-gcc
cd /home/pi/arduino-1.0.5fSE/hardware/tools/avr
rm -rf bin.gcc
sudo mv /usr/share/arduino /usr/share/arduino-1.0.1
cd /home/pi
sudo ln -s /home/pi/arduino-1.0.5fSE /usr/share/arduino
```

Download the ShrimpKey Software

With these commands you'll download the ShrimpKey software (the Arduino "Sketch") to the Raspberry Pi. In "Testing the Conductive Keyboard" on page 84 you used it to program the conductive keyboard:

```
wget http://fse.link/r10 -O ShrimpKey.zip
unzip ShrimpKey.zip
```

Copy the ShrimpKey sketch to the Arduino IDE's Sketchbook so you can open the sketch easily from the Arduino IDE later on:

```
mkdir /home/pi/sketchbook
mkdir /home/pi/sketchbook/ShrimpKey
cp /home/pi/ShrimpKey-master/ShrimpKey/*.* /home/pi/sketchbook/ShrimpKey
```

The Arduino IDE doesn't recognize the ShrimpUSB device by default, so you have to copy the hardware folder of the ShrimpKey software into the Arduino folder:

```
cp -r /home/pi/ShrimpKey-master/Arduino/hardware
    /home/pi/arduino-1.0.5fSE
```

The ShrimpKey uses a special library for the keyboard functions. This library has to be installed in the Arduino IDE (otherwise the sketch won't compile):

```
cp -r /home/pi/ShrimpKey-master/Arduino/libraries
    /home/pi/arduino-1.0.5fSE
```

Create a folder for the bootloader (this is explained in "Programming and Testing the Conductive Keyboard" on page 94):

```
mkdir /home/pi/bootloaders
```

Then copy the bootloader files (please type each command on one command line; there are only two spaces in each command, after `cp` and after `.hex`):

```
cp /home/pi/ShrimpKey-master/Arduino/hardware/ShrimpUSB/bootloaders/
ShrimpUSB/ShrimpUSB_rev1_atmega328p_16MHz.hex /home/pi/bootloaders

cp /home/pi/ShrimpKey-master/Arduino/hardware/ShrimpUSB/bootloaders/
ShrimpUSB/ShrimpUSB_rev1_atmega8_16MHz.hex /home/pi/bootloaders
```

Now create scripts for it with the text editor *Nano*:

```
cd /home/pi/bootloaders
nano atm328p.sh
```

Type the following in Nano (two lines):

```
sudo avrdude -c linuxspi -p atmega328p -P /dev/spidev0.0
    -U flash:w:ShrimpUSB_rev1_atmega328p_16MHz.hex

sudo avrdude -c linuxspi -p atmega328p -P /dev/spidev0.0 -U lfuse:w:0xD7:m
    -U hfuse:w:0xD0:m -U efuse:w:0x04:m
```

Press Ctrl-X, type Y, and press Enter. Type the following:

```
nano atm8.sh
```

Type the following in nano (two lines—the spaces are obvious):

```
sudo avrdude -c linuxspi -p atmega8 -P /dev/spidev0.0
    -U flash:w:ShrimpUSB_rev1_atmega8_16MHz.hex

sudo avrdude -c linuxspi -p atmega8 -P /dev/spidev0.0 -U lfuse:w:0x1F:m
    -U hfuse:w:0xC0:m -U efuse:w:0xFF:m
```

Press Ctrl-X, type Y, and press Enter.

Type the following commands to make the files executable:

```
chmod +x atm328p.sh
chmod +x atm8.sh
```

Add Configuration for USBasp Devices

The ShrimpKey doesn't use the default Arduino bootloader (a computer program that loads an operating system or some other software for the computer). Using the default bootloader would require you to use a USB-to-TTL converter (a hardware device) to program the chip. ShrimpKey instead uses software that simulates a USB chip, which allows a computer and the Raspberry Pi to detect the ShrimpKey as a regular USB device.

Normally, unknown USB devices can only be run with special permissions (with the root user account) on the Raspberry Pi. To make sure the ShrimpKey can be used under the default user account "pi" (or any other user), you have to tell the Raspberry Pi that it's safe to use USBasp devices with normal user accounts:

```
> sudo nano /etc/udev/rules.d/99-usbasp.rules
```

Type the following in nano (one line):

```
ATTRS{idVendor}=="16c0", ATTRS{idProduct}=="05dc", SUBSYSTEMS=="usb",
    ACTION=="add", MODE="0666", GROUP="plugdev"
```

Press Ctrl-X, type Y, and press Enter.

Run the following command to enable this change:

```
> sudo udevadm trigger --action=change
```

Install ScratchGPIO

ScratchGPIO is a special version of Scratch and isn't installed by default. You'll install it now, so that you can use it in "Project 2: Memory Game" on page 110:

```
> sudo wget http://fse.link/r11 -O isgh5.sh
> sudo bash isgh5.sh
```

If you're not logged in as the "pi" user, use this command:

```
> sudo bash isgh5.sh _yourusername_
```

Configure the Arduino IDE

The last thing that has to be done is configuring the Arduino IDE.

Start the graphical interface of the Raspberry Pi (if you're on the command prompt):

```
> startx
```

Run the Arduino IDE.

Go to Extra → Board and choose "ShrimpKey" (the one that matches your ATmega).

Go to Extra → Programmer and choose "USBasp."

Close the Arduino IDE.

You've now finished the manual installation of all of the necessary software. You can return to Chapter 2 to start making the conductive keyboard.

Index

S

T

U

V

W

About the Fonts and Images

All breadboard and stripboard images are made with Fritzing and edited with InkScape. These images are licensed under the Creative Commons Attribution-NonCommercial-ShareAlike 4.0 International License. The cover and body font is Benton Sans, the heading font is Serifa, and the code font is Bitstream Vera Sans Mono.

CPSIA information can
Printed in the USA
BVOW11s1534081114
374230BV00(